DATE DUE

MAR 2 4 1995			
MAR 1 2 1995			

THE ISLANDS SERIES

ALDERNEY

THE ISLANDS SERIES

*Published in the United States by Stackpole
†Published in the United States by David & Charles
‡The series is distributed in Australia by Wren

ALDERNEY

by VICTOR COYSH

DAVID & CHARLES

NEWTON ABBOT : NORTH POMFRET (VT)

Set in 11 on 13 point Baskerville
and printed in Great Britain
by Latimer Trend & Company Ltd Plymouth
for David & Charles (Holdings) Limited
South Devon House   Newton Abbot   Devon

For Leila,
who goes
with me

CONTENTS

ILLUSTRATIONS

ILLUSTRATIONS

Photographs not acknowledged above are from the author's collection

MAPS

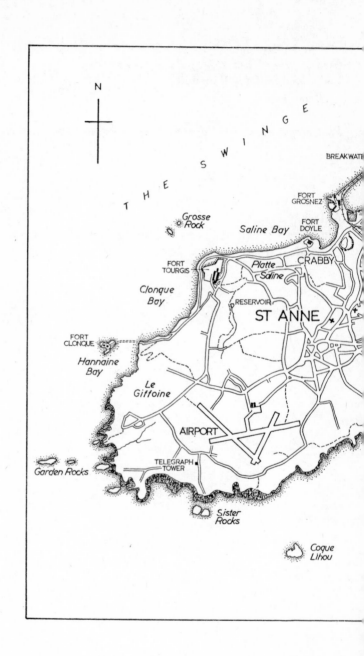

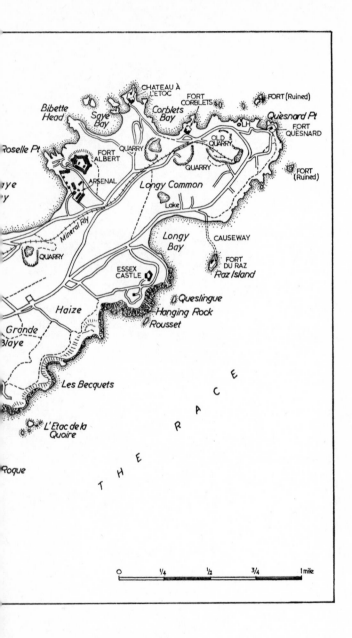

CHATEAU À
L'ETOC
FORT
CORBLETS
FORT (Ruined)
Bibette
Head
Saye
Bay
Corblets
Bay
Quèsnard Pt
LH
FORT
QUÈSNARD
Roselle Pt
QUARRY
FORT
ALBERT
OLD
QUARRY
ARSENAL
QUARRY
FORT
(Ruined)
aye
y
Longy Common
Mineral Rly
Lake
QUARRY
Longy
Bay
CAUSEWAY
ESSEX
CASTLE
FORT
DU RAZ
Raz Island
Haize
Queslingue
Hanging Rock
Rousset
Grande
Blaye
Les Becquets
L'Etac de la
Quoire
THE   RACE
Roque

0   ¼   ½   ¾   1 mile

1   THE ISLAND

IT stands, athwart the Channel tides, foursquare in a waste of waters—Alderney, the most northern of the Channel Islands, the nearest to England and France, but perhaps the least known of the archipelago. It is only about 9 miles from Normandy and, like its neighbours, its atmosphere still has a distinct Norman flavour.

Oblong in shape, Alderney lies between 49° 42' and 49° 44' N latitude and between 2° 9' and 2° 14' W longitude. It is approximately 3½ miles long, and 1½ miles wide at its broadest. Numerous isolated rocks and reefs surround it, among which Burhou and Les Casquets rank as islets.

The island comprises high land, rising to a maximum of 294ft above sea level at Le Rond But, east of the airport, and covering the western half, with low-lying territory occupying the eastern region. Much of the high ground is a plateau, and on the southern and western flanks steep cliffs add a touch of grandeur to the scene. The rest of Alderney rises only a few feet above sea level.

Guernsey lies some 24 miles south-west of Alderney, while the south coast of England is about 60 miles north. Approximately 8 miles west are Les Casquets, outposts of the Channel Islands, whose lighthouse is probably the most important in the English Channel.

While each of the Channel Islands has its own atmosphere, there is something about Alderney, hard to define, which makes it different from its neighbours. Beyond question the visitor is at once made to feel at home and, on his second and subsequent

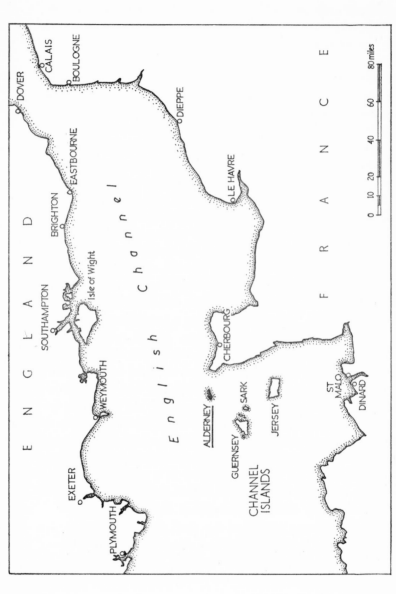

ALDERNEY WITH THE CHANNEL ISLANDS AND FRENCH COAST

visits, he is welcomed as an old friend. Added to this virtue are a democratic outlook and a cheerful regard for all and sundry.

Alderney-French, the dialect formerly freely spoken in the island, has completely disappeared, but placenames and certain surnames retain a French flavour, as does the appearance of St Anne, the diminutive capital. Its cobbled steets, old buildings, squares and lanes suggest the aspect of a small Norman town.

The port of Braye (which could well be deemed Alderney's tradesmen's entrance, the 'front door' being the airport) has character of a martial sort, owing to the almost overpowering presence of the breakwater, which, in the last century, was intended to turn Alderney into a naval base and harbour of refuge. The commercial element of the port is what matters today, yet in appearance the trading part of the harbour is insignificant compared with the mighty mole and frowning forts that still seem to mount guard over the spacious roadstead.

Like the other major Channel Islands, Alderney enjoys a certain degree of home rule, as it has done for centuries. Its population of 1,686 (the 1971 census total) is governed by the States of Alderney, a body comprising a president and twelve members, all of whom are elected by public suffrage. Two members represent their island in the Guernsey States of Deliberation, since Guernsey now has a hand in Alderney's affairs. There are some in the island, in fact, who resent Guernsey's control over Alderney's finances and major undertakings. They would like to see their island enjoy the independence of prewar years. But others, and they include Guernsey people, are happy with the present arrangements, which have worked admirably for nearly a quarter of a century. Alderney has remained solvent and relations between the two islands have been unswervingly friendly throughout the period.

Although Alderney is part of the Bailiwick of Guernsey (together with Sark, Herm and Jethou), there are many Guernsey people who have never been further than Braye or, at most, St Anne, when they have visited the island. A great many more have never been to Alderney at all. It cannot be said that Alderney folk are similarly ignorant of Guernsey, for business often compels them to visit it and, if they are travelling, sometimes this involves passing through the larger island.

Since 1945 the island has prospered exceedingly and its finances have maintained a healthy level, despite the large capital outlays necessary in its economic recovery. The United Kingdom has helped as well as Guernsey, but somehow Britain is regarded in a different light. Perhaps because Guernsey is a 'relative', as it were, and on something of the same level, Alderney regards the United Kingdom and Guernsey differently.

Those who travel to the island by air do so by Aurigny Air Services (see p 21) and this title keeps alive the old name of the island. Indeed, the French still style it thus and it is so referred to in the ceremonial parlance of the States of both Alderney and Guernsey, as the islands' parliaments are known. Precisely how the ancient *Aurigny* became the present 'Alderney' is something historians have been unable to explain convincingly. Some believe there is an association with Auderville, a Norman town just across the Race of Alderney. The 'ey' portion of the word suggests 'island', but exactly how 'Alder' came into it is likely to remain as great a puzzle as is the origin of the other Channel Islands' names.

Another name which the visitor may observe is *Riduna*, and some believe this was the Roman title of the island. According to the *Itinerary* of Antonius, Guernsey was called *Sarnia*, Jersey *Caesarea*, Sark *Sargia* and Alderney *Riduna*, but historians are sceptical, declaring that the names refer to other islands entirely.

Turning to the present, in some respects the cost of living is rather higher in Alderney than in the United Kingdom, especially food prices, but rates and taxation, together with the prices of liquor, tobacco, perfume and the absence of value added tax, more than offset the disadvantages. To buy a house, though, is not a cheap undertaking, for even the most humble

Page 17 Alderney, seen from the east, is almost severed in the middle by Longy Bay, to the left, and Braye Bay. Offshore forts on the left and right are as conspicuous as the breakwater, top right. The island's roads show up white, like the town of St Anne at the top of the photograph

Page 18 This aerial view shows Mannez Lighthouse in the centre and the open terrain at the island's eastern end. Left is Fort Quèsnard and near it is the isolated Fort Houmet Herbé. Few buildings occupy this part of Alderney, whose quarries are seen on the right

cottage today commands a high price. Expensive, too, is the cost of electricity, because of the relatively small nature of the undertaking.

Another small enterprise is the *Alderney Journal*, a lively periodical, published fortnightly. It succeeded the *Alderney Review*. The daily newspaper is the *Guernsey Evening Press*, though unfortunately it arrives on the day following publication. Radio and television, of course, are in practically every home.

Other forms of recreation include football. There is an inter-insular competition, held annually for the Muratti Vase, and football teams from Jersey, Guernsey and Alderney compete. It speaks much for Alderney's sporting spirit that, although she has only won the trophy once, in the 1920s, her team still competes.

Besides football, there is cricket, which is played on Butes, the recreation ground in St Anne, from which a noble panorama of sea and land can be enjoyed. Golf, too, is important, thanks to the excellent nine-hole course, opened in 1970, and replacing a links formerly used by the garrison, which was destroyed during the German occupation.

Other local sport includes shooting, fishing (excellent opportunities exist all around the coast), sailing (there is a thriving sailing club, with headquarters at Braye) and indoor activities such as bingo, darts, bridge and whist. Films are also shown in the island periodically, and occasional theatrical entertainments are given.

Church services are fairly well attended and at the great festivals congregations are large. Church social events receive generous support, particularly St Anne's Church fête, normally opened by the Lieutenant-Governor.

In winter dinners are popular, whether they be of a private nature or gatherings of old folk or organised bodies, such as the Royal British Legion, whose annual dinner at the Sea View Hotel is one of the outstanding events of the island calendar.

Islanders also support the efforts of the local dramatic society and during Alderney Week the various attractions, including a torchlight procession, cavalcade, play and beach competitions, delight visitors and islanders alike. The Alderney Orchestra unfortunately belongs to the past, like the band of the Salvation Army, but there are occasional visiting artistes.

Life need never be dull, for culture is by no means forgotten. The public library is particularly good. The Alderney Society's museum plays a major role in the island's intellectual life, endeavouring to preserve what is best in Alderney; and the work of the Society (see p 40) leaves its imprint on several elements of the community.

2   COMMUNICATIONS

ALDERNEY is easily accessible from England and the Continent at all times of the year. The most common approach is by air. Flights between the islands often take less than 15min, the same time as the journey between Cherbourg and Alderney, and one may reach the island from Southampton in about 45min.

During the summer Aurigny Air Services operate as many as a dozen daily flights from Guernsey to Alderney, as well as four flights per day from Jersey and seven from Southampton. There are also daily flights between Alderney and Cherbourg. Reservations can be made at Southampton, Guernsey and Jersey airports, at the company's offices at the South Esplanade, Guernsey, and the Weighbridge, Jersey, as well as at the Alderney airport or the town office in St Anne. The firm's booking office at Cherbourg is at Deshayes, Agence Maritime, 20 Bis a Rossel.

Because of its convenience, the vast majority of visitors use air transport, but another popular method of travel is the hydrofoil, an exciting experience. Usually the trip between Alderney and Guernsey takes less than an hour and it has the advantage of using Guernsey's capital, St Peter Port, rather than the airport. The service, by Condor Hydrofoils, starts at the end of March and ends at the close of September. There are sailings between Guernsey and Alderney (with connections with St Malo and Jersey) on Tuesdays and Thursdays. Tickets are obtainable from Condor Ltd, Passenger Department, North

Pier Steps, Guernsey, or from Commodore Shipping Services Ltd, 28 Conway Street, Jersey. On a day excursion to Alderney the visitor is allowed about 4hr ashore. A day excursion by air gives one longer in Alderney and, of course, the same advantage applies to those in the smaller island who fancy a day in the larger.

A few passengers may travel by the freighter that runs between Guernsey and Alderney, but the service is relatively infrequent. During the summer, it is sometimes possible to go to Alderney by excursion vessel from Guernsey, and to have several hours ashore, but, unhappily, gone are the days when this was possible by Weymouth excursion steamer.

A considerable number of visitors to Alderney reach there by their own means. Some travel by private aircraft and some by charter plane, but the majority go by yacht. Alderney is a comfortable day's sail from several south of England ports and its proximity to Cherbourg and St Peter Port is a further advantage to the yachtsman, who also uses Alderney as a port of call en route for the other Channel Islands or perhaps Brittany.

Mails and gales

Thanks to the good air service, posts are frequent, a state of affairs far removed from the days when the island's sole link with the outside world was a small steamer which called twice a week, weather permitting. Even then she only operated to and from Guernsey. Nowadays, weather can affect flying, and when this happens the post is necessarily delayed, like the passengers. Strong winds are sometimes responsible for such interruptions, but fog is the worst enemy. To wait for a flight sometimes entails hours of delay, but in Alderney it is usually possible to stay at one's hotel until notification is received that conditions have improved.

It is curious to note that, whereas today one can often fly between Alderney and Guernsey in as little as 10min, not so long ago it took nearly 24hr to cover the distance by sea. This was not in the days of sail, but when fog was responsible. In 1913 the *Serk*, replacing the regular *Courier* for the time being, took 27hr to do the journey. Most of the time the steamer was at anchor off south-west Alderney. Again, in 1953, the MV *Island Commodore* (later renamed *Ile de Serk*) was 21hr on passage, again due to deplorable conditions.

Although the elements are as formidable as ever, the amount of Alderney shipping they affect is far less than hitherto. Today, no stone or cattle is exported, although gravel is shipped from Braye and sometimes sand. Fuel is imported and heavy goods are carried in the regular freighter.

INTERNAL SERVICES

The visitor arriving in Alderney will soon discover that internal communications are not wanting. Taxis await him at the airport or the harbour, and they are both fast and comfortable. They play a big part in island life, for although many residents own cars, there are occasions when taxis are preferable.

In Alderney it is easy to hire a car, a scooter, a pedal cycle and even a horse! The roads are good and the volume of traffic, although greater than it was a decade ago, is not enough to make travelling disagreeable. In St Anne one may hire vehicles, while horses are available from Verdun Farm, on the Longy road and barely $\frac{1}{2}$ mile from the town centre. Because of the increasing number of vehicles in the island, a few of the streets of St Anne have been made one-way thoroughfares and there are parking restrictions in the principal shopping centre.

Large, comfortable and reasonably priced buses ply from Marais Square, St Anne, down to Braye and thence along the coast road eastwards. The buses serve the bays of Braye, Saye, Corblets and Longy, and they will stop practically anywhere. It is a good idea to take the bus in one direction and walk back, or vice versa. The bus services are fairly frequent and they operate during the summer months, including Sundays. Out of

23

season there are school buses only, on which adults may travel.

Commercial vehicles are numerous and, nowadays, all are mechanically driven. The state of the roads has improved as a result of this upsurge in the vehicle 'population', though several in the heart of the town retain their picturesque paving stones. If their contours are a little rugged, this helps to reduce speed where it is least needed.

Telephones

Alderney's telephone system is controlled by the States of Guernsey Telephone Council and comprises nearly 700 sub-scribers. Radio telephony is employed and the charges, like Guernsey's, are substantially lower than those of the United Kingdom—hence the undoubted popularity of the system, which is automatic. The original exchange, installed in 1949, was manually operated and was replaced by the present one in 1970.

3   ROCKS, FAUNA AND FLORA

'THE Channel Islands are portions of France which have fallen into the sea and been picked up by England,' wrote Victor Hugo in his romance *The Toilers of the Sea*, in the last century. His words have more relevance to geology than to history, for in 1066 it was England that was 'picked up' by Normandy, of which the Channel Islands then formed a part.

THE REMOTE PAST

Going into the past beyond written history, beyond even the evidences of archaeology, fluctuations of sea level during the past 1½–2 million years gave to Alderney, as to the other Channel Islands, an alternating succession of island and Continental existences. As sea levels rose and fell in response to the extreme climatic changes, so Alderney was now an island much as at present, now one with the mainland of Europe.

Further back still the island's identity becomes one with the geological history of 'the great Armorican province of North-West France', as Dr J. T. Renouf describes the region in his recent synthesis for the Alderney Museum. The first recognisable event in Alderney's geological story is found in dark patches, or enclaves, as the geologist would have them, which can be seen by anyone examining the grey igneous rocks of Clonque. These rocks are called granodiorites and are now dated to more than 2,000 million years ago, a fact indicating that the dark enclaves are older still, for they were already solid when caught up by the grandiorite in its molten state deep in the earth.

Both enclaves and grandiorite belong to what has been termed the Western Igneous Complex. In many places the grandiorite has a distinct foliation, that is, a set of more or less regular planar weaknesses resulting from an episode of intense

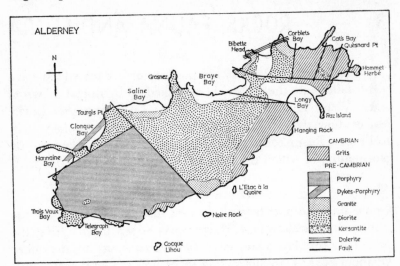

pressure and heating (metamorphism) at some unknown time after its initial formation. Giving bright colour to what would otherwise be rather drab rocks, a number of reddish-purple dykes seam the shoreward cliffs of Clonque and indeed most of the other cliffs fringing the western coasts of the island. These minor intrusions also predate the rocks of the Central Diorite Complex, next to be described.

Central Diorite Complex

These oldest of Alderney rocks occupy the high ground in the western part of the island; between them and the sandstones of the east lies the great wedge of the Central Diorite Complex, which is bounded by the rough triangle formed of Fort Doyle, Corblets Bay and La Cachalière. The diorites of the central complex lack the metamorphic textures of those found in the west and are undoubtedly much younger, though only the very latest event associated with this complex, the intrusion of the Bibette Head granite, has been dated. The granite is of Cambrian age and was intruded some 500–600 million years ago, this figure being taken as the approximate upper age limit to the several rock types that constitute the complex as a whole.

Small outcrops of picrite, gabbro and layered and orbicular diorites in the vicinity of Roselle Point, below Fort Albert, give variety to the main diorite mass. The fresh diorite is best examined in the old quarries of La Cachalière. The most fascinating of the minor rock types mentioned is the exposure of orbicular diorites near Roselle Point. As Dr Renouf points out, they have a fame beyond the confines of Alderney, 'for they are among the finest examples known of the geological phenomenon of mineral diffusion to form concentric banding'. Many standard books of reference make mention of these Alderney rocks. A typical orb has an alternation of plagioclase and hornblende-rich zones.

The Bibette Head granite is noted for its dark enclaves, often spotted with pale feldspar crystals. These are very abundant locally. Dykes of dark grey lamprophyre (this rock weathers to an ochreish yellow-brown above the reach of the tide) and dolerite dykes, also dark grey, nearly cut Bibette Head off from the coast. The lamprophyres here are rather similar to the one or two examples that cut the Western Igneous Complex, and are probably of the same age.

Alderney Sandstone

While the time interval separating the formation of the Western and Central Complexes is of the order of more than 1,000 million years, the deposition of the Alderney sandstone probably followed the intrusion of the Bibette Head granite by a mere 50–100 million years at most. A mountainous landscape was raised in the wake of the wider disturbances, of which the intrusion was a part, and this was denuded to provide the

27

gravelly sand built up in the shallow turbulent waters to form the Alderney sandstone.

The red and green colours that are so striking a feature of the sandstones represent original variations in the sedimentation processes, probably related to climate. Cross bedding, that is, lines running obliquely across between the main bedding planes, is extremely abundant. Some ripple marks and other structures indicative of a water-lain origin are also to be found. A possible fossil find was made on the beach near the eastern end of Braye Bay, where a boulder of pipe rock shows clear evidence of the activity of boring organisms. The boulder closely resembles similar rocks from the north of Scotland. The small islands of Burhou, the Ortac rock and the Les Casquets, together with an outcrop at Omonville-la-Rogue, east of Cap de la Hague, testify to the once considerable extent of the Alderney sandstone.

The Great Ice Age

Another great interval of some 400–500 million years separates the deposition of the Alderney sandstone from the incomplete sequence of poorly consolidated material representing the 'soft' rocks of the Pleistocene, or Great Ice Age. In Alderney only the closing stages of the Pleistocene are represented, in the raised beaches and periglacial solifluction deposits, known as 'head'.

The changes in sea level previously referred to are recorded in similar successions of deposits occurring at two different heights above Ordnance Datum (OD), each representing a different cycle of ice retreat and ice advance away north in England and across northern Europe. Some 150m south-east of Quèsnard lighthouse a 5m vertical section at 18m above OD is represented by large sea-worn boulders at the base, passing upward into angular rock fragments in an orange brown matrix—the head. The beach is some 120,000 years old and preceded the last ice advance of the Riss Glaciation, when the head was formed. Following the Riss there was a warmer period, with consequent rise in sea level, though it did not reach the 18m of the Quèsnard exposure.

At this time of high sea level, called the Eemian, the sea stabilised at about the highwater mark of present ordinary spring tides, and around the island there are many raised beaches and wave-cut notches at this level, which is 8m above OD. Capping these raised beaches is often several metres of head, formed during the last or Würm Glaciation, between 60,000 and 20,000 years ago. A very fine section is to be seen in the rubbly vertical cliffs just west of Clonque causeway.

Later deposits belong to the last 10,000 years and are mainly coastal dune formations, such as are found at Longy Bay. Weathering of the rocks has shaped the cliffs into their present forms and has also developed a soil, though thin in many parts, over most of the island. Changes within this period are recorded by such events as the overwhelming of the Iron Age site near the Nunnery by blown sand.

The island offers a number of small mineral finds, such as manganese dendrites and other coatings to fault planes, haematite, iron pyrites, epidote and hornblende, but generally speaking it is the rock types rather than their individual minerals that are the most interesting.

MAMMALS

The mammals of the island are neither varied nor numerous. Rats and mice are to be found in and around buildings, and according to Ansted & Latham's *The Channel Islands* (1862) the black rat was once common. Rabbits abound on the cliffs and commonland. During and immediately after the German occupation the presence of rats was very apparent, while rabbits seem to have led an unchecked existence for over 5 years. However, occasional outbreaks of myxomatosis have

29

greatly reduced the latter, though they appear to be becoming immune to this terrible disease.

Of particular interest is the Alderney shrew (*Crocidura russula*), which is also found in Guernsey and Herm, but not in Britain. A different species is the white-toothed shrew, found in Jersey and Sark.

Horses are more plentiful than they were since the days of the occupation, when the Germans made great use of them. Today they are ridden only, the internal combustion engine having almost completely ousted the horse-drawn vehicle in Alderney. Donkeys are rare, though a very few may be seen, and cats and dogs are ubiquitous.

Contrary to common belief, the Alderney breed of cattle never existed. An authority on the subject of Channel Islands cattle, the American E. Parmiloe Prentice, writing in the *Transactions* of La Société Guernesiaise for 1946, stated that the first cows imported into England from the Channel Islands were from Alderney and the word 'Alderney' became synonymous with 'Guernsey' and 'Jersey'. He wrote that Guernsey farmers were glad to sell their cattle as Alderneys—this was before the days of rigid herd recording and export regulations—and later Alderney cattle were registered in the Guernsey Herd Book 'and the name "Alderney" is no longer used as a breed name'.

It was said that the Alderney cow was smaller than the Guernsey, but in fact it was of the same breed. Before the Germans landed there, nearly all the cattle in Alderney were shipped to Guernsey, and after liberation Alderney was stocked with animals from Guernsey.

Alderney seems to be lacking in reptiles, according to *The History of Alderney* by E. A. Martin, written in 1810, and has no foxes, hares, stoats, weasels or hedgehogs.

MARINE ZOOLOGY

Alderney has fifteen species of anemones, including the dahlia (*Tealia crassicornis*), the opelet (*Anthea cereus*) and the cup coral (*Balanophyllia regia*); the daisy (*Sagartia bellis*) is more common in Guernsey. Although sponges exist, they do not flourish in great numbers. Apart from a sea cucumber and a sand star, echinoderms are not plentiful. Univalves, bivalves and nudibranches are similar to those of Guernsey.

Seaweeds abound and such beaches as Telegraph Bay, Clonque, Cats Bay (near Mannez lighthouse) and the shore between Fort Houmet Herbé and Longy provide happy hunting grounds for the marine biologist. Any shingle beach will yield plentiful varieties of shell, while such marine creatures as limpets, topshells and, most notable of all, ormers, are easily found, especially at a low spring tide.

The ormer (*Haliotis tuberculata*) is the Channel Islands' largest mollusc, and it is not found in the British Isles. Ear-shaped, it can attain a length of 4in and a width of 2–3in. Its single shell is beautifully lined with mother-of-pearl. It is to be found clinging to rocks and stones below the level of low-water spring tides and its removal depends upon the speed of the operator wielding an ormer hook. This fish may not be taken at all times of the year—there is a close season during the summer months. The ormer is considered to be a great delicacy and the shell is prized as an ornament.

Other shellfish found in Alderney waters include crabs (particularly 'spiders' and *chancres*), lobsters and crawfish. Such marine creatures as the octopus are sometimes encountered, while edible fish caught include conger, mackerel, pollack, garfish, ray, turbot and bream. In rock pools one commonly sees the *cabou* (blenny), as well as prawns, shrimps and many other specimens.

BUTTERFLIES AND MOTHS

Alderney is not noted for any specialities, though it must be confessed that not a great deal of work has been done there by lepidopterists. The painted lady is common in the island, as in its neighbours, and so are the Glanville fritillary, the clouded yellow and the meadow brown. Occasionally such rarities as the Queen of Spain fritillary and the Camberwell beauty are found.

Moths abound, though the death's head is less common than hitherto. More plentiful are the privet hawk, convolvulus hawk and humming-bird hawk. The tiger moth also may be met with.

BIRDS

According to Peter Conder's *List of the birds of Alderney, 1972*, published by the Alderney Society, there are 196 species to be found there. Obviously, the majority are to be seen around the cliffs and in the open country, though the gardens of St Anne attract tits, finches and other engaging birds. Rarities like the hoopoe, bee-eater and golden oriole have been reported and the sportsman may well put up snipe or woodcock in winter. It is curious that the magpie, so common in the other Channel Islands, is a rarity in Alderney, yet it was plentiful in the last century.

Waders are common, especially oystercatchers, and a few duck may be encountered, though, like geese, they are not found in much variety. Curlew are seen and the kestrel is listed by Conder as a 'common resident', like the cuckoo in summer and the skylark, among many others, Rock pipits haunt the shore and the carrion crow is equally common. So are the blackbird, robin, thrush, starling, chiffchaff, lapwing, swallow and house martin. The Kentish plover breeds in small numbers, to the delight of the naturalist, and such colourful birds as the kingfisher and great spotted woodpecker may be seen by the fortunate. The heron is an occasional visitor, like the whimbrel and the short-eared owl. The wryneck, once common, is less often seen today.

Seabirds

Alderney is less famed for its land birds than for those of the sea. Of these, the gannet is king, beyond dispute (see under Les Etacs and Ortac, rocks where it breeds, pp 34 and 37). Divers are rare, but the fulmar, once a scarcity, has been observed since 1962 fairly often. Sometimes various types of shearwater visit Alderney, as do the terns.

Greater and lesser black-backed gulls, herring gulls, black-headed gulls, shags and cormorants are very frequently encountered, and on Burhou it is delightful to observe the antics of puffins and, at night, to hear the sounds of storm petrels. On outlying stacks razorbills and guillemots breed.

The naturalist R. M. Lockley, writing in *British Birds* (Vol XXXIX), reported that the presence of gannets in Alderney waters was first observed by E. Quinain, an island fisherman, in the summer of 1945. They had been seen in very small numbers before, but in nothing like the quantity observed in that year. They were on Ortac and covered the rock. He saw smaller numbers on Les Etacs. A further reference in *British Birds* (Vol XLI) stated that Major J. A. A. Wallace, commanding the Machine Gun Training Centre in Alderney in June 1940, went with others to Ortac looking for kittiwakes and found a gannet on its nest, on top of the stack. He noticed an egg after the bird had flown off. This was the first record of the gannet breeding in the Channel Islands. This officer landed at Les Etacs at the same period, but found no gannets there.

The same journal also reported another new gannetry on Rouzic, one of the Sept Iles, off northern Brittany. It appears this colony was established at about the same time as that on Ortac. The reason why the Alderney gannetries came into

33

being has never been explained very convincingly, though it has been suggested that, early in the war, the island of Grassholm, off south-west Wales, which formerly hosted gannets, was used as a bombing range and the birds resettled farther south, blissfully unaware that they were in enemy territory and in even more warlike surroundings!

Les Etacs

The Garden Rocks, as many style these fine crags, are easily climbed, though what is less simple is finding a boatman prepared to take one there. The rocks, near the western end of Alderney, stand 128ft above sea level and are in two groups. Most of the birds are on the larger and more westerly islet, where there is a little soil and vegetation, chiefly comprising sea-beet and sea-pinks. Les Etacs stand very close to the cliffs of Trois Vaux Bay.

Here are some of the 2,000 pairs of gannets (according to the latest count), the rest being on Ortac. Roderick Dobson, in his excellent *Birds of the Channel Islands*, repeated R. M. Lockley's statement that in 1946 there were 200 pairs on Les Etacs and 250 pairs on Ortac. There were then 190 nests on Les Etacs, mostly on the central northerly slope of the larger rock and a further 100 pairs on a pinnacle in the eastern group. Thus, numbers have greatly increased since then and, as the gannetries became overpopulated, some birds at long last moved to smaller stacks in the group. Yet they ignore other offshore stacks, such as Cocq Lihou, Les Nannels and, for that matter, the Casquets and Burhou.

Gannets are the largest British seabirds, with a wingspan of 6ft and a formidable beak, but they are strangely docile at close quarters. They are noisy enough in their indignation at man's intrusion on their domain, yet they suffer him near their nests (even if they contain chicks or eggs) and actually permit themselves to be picked up and ringed. The smell of a gannetry is strongly acrid, though one becomes used to it in time.

Page 35 (*above*) The wrecked sailing ship *Liverpool*, ashore off Fort Hommeaux Florains in 1902. The world's largest sailing ship at the time, it became a total loss, though no lives were lost; (*below*) the steamer *Courier* at Braye. For many years she was the link between Alderney, Guernsey and France. Despite her sinking off Jethou in 1906 and innumerable buffetings, this gallant ship braved the elements in island waters and further afield until she was broken up in 1951 at the age of 68 years

S. "Courier," leaving Alderney.

Page 36 (*above*) Les Casquets in 1844, when three lighthouses stood upon the rocks. Today the towers remain, but a single powerful light is sufficient. The rocks have been lighted since early in the eighteenth century; (*below*) Mannez lighthouse, standing at the eastern end of Alderney, built in 1912 and shown here soon after its completion

The nests are built in terraces, advantage being taken of the islet's rocky terrain. They are made chiefly of seaweed and seem to be built up every spring, when the birds return from wintering off the African coast. Sometimes nests have flotsam in their construction. Gannets are not averse to neighbours breeding in close proximity and at Les Etacs guillemots are apparently made welcome in this noisy, crowded and confined space. The din and odour can be appreciated (if that be the word) from La Giffoine, the point on the Alderney coast from where the gannetry can be observed with ease, preferably with binoculars.

To watch a gannet diving for fish is most impressive. It descends with a headlong rush, continuing the plunge under water and then immediately surfacing, triumphant and ready for more 'dive-bombing'. The birds are often seen from Alderney returning from fishing, flying low and in formations of from three to six. Landing on Les Etacs is not easy for the birds, so thronged are the ledges, but it is accomplished almost continuously and, it seems, with complete success.

Writing on the ways of gannets in the Alderney Society's *Bulletin* for January 1972, Peter Conder noticed that those returning from the Cap de la Hague direction flew down Alderney's north-west side rather than the direct south-east coast, 'those flying to Ortac tending to keep as far out as Burhou', he wrote, adding that he had no idea why they did this, nor why they should always return to their colonies by the east or west routes.

Ortac

This massive rock, which has no vegetation, stands about 8oft above the sea and its crown is whitened by the birds' presence and their droppings. It is said to resemble a haystack in shape and it has always been something of a hazard to shipping. Its name may mean 'the large rock at the edge' ('or' meaning edge and 'etac' a large rock) and it stands on the end of a chain that includes Burhou and the Casquets. Both Dobson and Lockley, writing in the 1960s, estimated its gannet population at about 500 pairs. Beyond this number, they thought, it would be impossible to go, because of the rock's limitations, although some ornithologists declare the figure in the 1970s to be about 600 pairs. Possibly even this is an underestimate.

There is usually a swell around the rock, but on a calm day it is relatively easy to land. The climb to the top is simple. Its occupants, like those of Les Etacs, suffer intruders with noise and commotion, but normally show no real animosity. Again, a few guillemots share quarters with the much larger birds. Kittiwakes used to nest there as well, though they seem to have forsaken the rock now.

The view from the top is vast and the prospect of the Casquets, Burhou and Alderney is a novel one. Folklore records that a great cave known as 'The Oven' exists at Ortac, and that when intruders arrive, the kittiwakes hurl their eggs on to its floor. A visit to Ortac reveals nothing of the kind. Nevertheless, the birds formerly nested on ledges and possibly they pushed the eggs 'overboard' at the approach of strangers. Perhaps the gannets' presence has driven the kittiwakes away, for the great birds are often in residence from as early as January to as late as the end of October, thus giving other seafowl little opportunity to breed in peace.

The people of the Guernsey Bailiwick are regularly reminded of the Alderney gannet by the Guernsey penny, on which it is shown in flight.

FLORA

Excellent work has been done in this field by Frances Le Sueur (of Jersey) & David McLintock in *A Check List of the Flowering Plants and Ferns wild on Alderney and its off-islets*, published by La Société Guernesiaise in 1964. They agree with the Guernsey naturalist E. D. Marquand, who, in 1899, maintained that a single day's botanising in Alderney would yield better results than a similar period in any other of the Channel Islands.

These contemporary botanists feel that such upheavals as the building of the breakwater and forts in the last century and the German occupation played a part in the botany of the island, causing disturbance and neglect respectively. The relative scarcity of tourists, as compared with the other Islands, has also done much to save the flora of Alderney.

In 1901 there were 519 species of plant, and the number in 1964 was 682. There might have been more had Longy Pond, on the Common, retained its water and if building operations between 1960 and 1970 had been less intense. On Burhou, which is uninhabited, the list was as high as 45 and, oddly enough, there is a marked contrast between Burhou proper and Little Burhou, although only a few yards of beach (covered at high tide) separate them. It is perhaps significant that Burhou's list is two and a half times as numerous as that of 1900, although the botanists advance no reason for this increase.

Purple spurge (*Euphorbia Peplis*) is found in Alderney, but nowhere else in the British Isles. Another rarity is *Romulea Columnae*, sometimes called the sand crocus. Yet another is the the *Polycarpon Tetraphyllum*. It is relatively common in Alderney, like the broom rape, *Orobanche* and the *Orobanche Minorvar Flava*.

The ordinary visitor will delight in Alderney's sheets of gorse, broom and heather. Spring brings forth primroses and celandines, bluebells and violets, speedwell and campion. Foxgloves add dignity to the floral picture, as does the orchis, which can be found with little effort. Vipers bugloss, thrift, daisies, buttercups and blackthorn adorn the contours of Alderney in due season, among other plants, not forgetting ivy which, too often, infests trees with its tendrils.

Scrubland on the cliffs and other open spaces (especially along the railway track between Mannez and Braye) includes the inevitable brambles, which yield blackberries in plenty. Despite the winds which so often blow over Alderney, its gardens are gay with flowers, thanks, in many instances, to the stone walls that serve as windbreaks.

Although trees are not on the whole very plentiful, in the valleys about St Anne they are quite abundant. Near the church is the Terrace, a somewhat neglected public garden, which is well wooded, like the Valley beside it, running down to Crabby. More trees decorate the scene at Le Petit Val, extending from Royal Connaught Square to Platte Saline. Valongis, running from Les Rochers down to Braye, has some trees, but there are many more in Val Reuters, or the Water Lane, which meanders downhill from the top of Le Val to Newtown.

The Alderney Society is interested in tree planting and tree preservation, an interest shared by the States of the island. At Val du Sud, on the south cliffs, their efforts may be appreciated, and a fringe of trees along the Lower Road, skirting Braye Bay, softens the aspect quite effectively.

Particularly well wooded is 'Essex Glen', which begins near the road junction opposite the golf clubhouse on the Longy road and proceeds parallel with it to Longy Bay. This shallow valley had much vegetation on either hand, and makes an agreeable alternative to the somewhat prosaic main road.

Generally speaking, sunshine, rainfall and temperature do not greatly differ from those of Guernsey and, compared with England, winters are mild in Alderney, although the force of the wind, whether from the north-east or the south-west, can certainly be felt. Yet the air is bracing and the climate is healthy, something borne out by the longevity of many Alderney folk.

Average rainfall is approximately 32in, while the sunshine in a year amounts to about 1,760hr.

H. D. Inglis, in his work, *The Channel Islands*, published in 1835, was critical of Alderney's climate. The absence of wood, he wrote, left the course of the winds unbroken and the only shelter was the stone walls, 'here and there used as enclosures', instead of hedges, which is still the case, incidentally. It was not to be expected, therefore, that 'the more delicate fruits and flowers will attain, in Alderney, the perfection which they reach in the sister islands', unless they were grown under glass. Yet one has only to attend a fruit and flower show in modern Alderney to witness a spectacle every bit as good as one may see elsewhere. Sheltered gardens, the use of glass and the presence of some trees, at least, conspire to make the island as healthy a place for the produce of its soil as it is for those who reside there.

BECAUSE of the number of Neolithic and Iron Age remains found in the island, there is good reason to suppose that it was once a prehistoric necropolis. Alderney lies only about 9 miles from the coast of Le Cotentin, and one may well assume that primitive man, regarding offshore islands as places of sanctity, honoured his illustrious dead by ferrying them over the Race for interment.

Unhappily, few of their tombs survive in Alderney. The building of forts, a light railway and, no doubt, access roads in the nineteenth century caused the destruction of many a prehistoric tomb, menhir or stone circle, and probably their material was often used in the construction of the works. Fortunately, while this work of destruction was in progress, the antiquarian family of Lukis, of Guernsey, visited Alderney, and while they were quite unable to save the remains themselves, they were able to record their existence and positions; some 'finds' were removed to the Lukis Museum and also to the Guille-Allès Library Museum, both in Guernsey. Also housed in the family museum was the massive *Collectionea Antiqua*, six volumes comprising 400 pages of manuscript, numerous water-colour drawings and pen and ink sketches of the Channel Islands' prehistoric remains, the work of Frederick Corbin Lukis, the eminent Victorian antiquary.

PREHISTORIC REMAINS

The prehistoric structures were chiefly located at Longy, the flat region on the south-east coast, overlooking the Race. On the slope above the bay stood Les Pourciaux, a name associated with pigs and perhaps so called because the stones resembled swine when viewed from a distance. This monument, among others, was examined by Francis Lukis.

It comprised three sections, and excavations disclosed pottery fragments and both human and animal bones. An unusual feature of the northern structure was three small cells, roofed with stone slabs and containing human bones, but no pottery. They stood on a stone pavement. Les Pourciaux remained intact until the German occupation, when two of the sections were destroyed. A capstone and three uprights of the survivor exist in the garden of a residence at Longy, but unfortunately a German bunker covers the rest of the monument.

In the north-west stands Fort Tourgis, and nearby is a cist, comprising two props, 6–7ft long and standing about 2ft 6in above its floor, lying parallel with one another and 2ft 6in apart. Its capstone is triangular in section. The cist stands in a furze clearing just off the road as it runs down to the fort. When John Lukis excavated it, he found no bones or pottery, probably because the cist's existence had been known for centuries. It is the best preserved megalith in Alderney.

North of the Longy road and on high ground overlooking Braye Bay is the region of Les Rochers, now dominated by a lofty television mast. Hereabouts one finds many blocks of stone—hence the name—and these some believe to be of prehistoric origin. They are not part of outcrops. Probably many more were used in the building of the relatively modern walls in the vicinity.

In Longy Bay is Raz Island—'Isle of the Race'—and here in the nineteenth century a fort was built, linked by causeway with the coast. During excavation work a prehistoric burial ground was found. A number of small mounds covered graves and Francis Lukis saw these before their destruction. The graves mostly comprised rough flat stones, usually covered by capstones and, deeper than these, human remains and elegant bronze daggers were found. Possibly the mounds were raised over a Bronze Age cemetery.

While building work progressed, a pottery vessel, a bronze ring and some coins were found, but the coins were thrown away by the finder. However, Francis Lukis, on hearing of this, interviewed the workman and obtained from him the vessel, which was displayed in the Lukis Museum. It was of Roman make and was almost completely filled with burnt bone.

In the same year, 1853, what Kendrick deemed to have been a Bronze Age interment was discovered at Raz. In a cist was a skeleton 'with legs folded over and doubled up', according to H. P. Le Mesurier, who wrote to Lukis on the matter. On the ribs was a bronze dagger, clasped in the skeleton's folded hands. The dagger, unfortunately, was not acquired by the Lukis Museum, having been 'secured' by an army officer at the time of discovery. The cist's capstone was found lying on the side when Le Mesurier saw it.

When a fort was under construction at Château à l'Etoc, on the north coast, another early burial ground was discovered. This was again in 1853, but when Francis Lukis arrived on the scene its destruction was almost complete. The men engaged on building work were continually finding human remains and pottery, and these were thrown over the walls on to the rocks. Lukis stated that 'the whole surface of the hill was filled with the depositions of the dead', but no plan was made of the precise area of the cemetery.

Among the megaliths destroyed in the nineteenth century was Alderney's only known menhir, La Pierre du Vilain, which stood near Longy pond. Not far off, close to Essex Castle, Francis Lukis excavated a barrow and a cist, discovering pottery, flint chips and an 'anvil stone'. Today there is no trace of these remains, and a stone alignment near the lighthouse has also gone—a victim of the German occupation.

Near Telegraph Tower, on the south-western cliffs, there

stands La Houguette de la Taillie, a tumulus excavated by
Francis Lukis in 1853. Within the mound he discovered the
remains of a cist, flint chips, potsherds and an 'anvil stone'.
Hereabouts stood a beacon, on a site known as Le Béguine,
which would have been kindled at the approach of an enemy.

The Iron Age Site

In 1968, while work was in progress on the new golf course, an
earth remover turned up pottery in the area of Les Huguettes,
between Les Rochers and Longy Bay. Peter Arnold, the potter,
happened to be there at the time and he at once identified the
finds as prehistoric. Kenneth Wilson, the archaeologist, suspected they were of late Bronze or Early Iron Age date and,
with his wife and others, thoroughly examined the site, thanks
to the cooperation of the Alderney Golf Club authorities.

Excavation revealed that the earth remover had disturbed
about two-thirds of a hut, 'measuring about 20 feet long and 16
feet from one side of it to where it disappeared into the un-
touched area', to quote from Wilson's report. He found traces
of a wall and 'a large area of the hard-packed floor lay beneath
a layer of burnt clay daub'. A hearth and balls of clay were discovered and it was supposed that the hut had caught fire, the
roof falling in and crushing the pots on the floor. Later, Mrs
Wilson restored several of them and they are now on view in the
museum.

Wilson thought the hut had been a workshop rather than a
residence. He found no metal at Les Huguettes and declared:
'It is possible that many of the offshore islands could not obtain
metal and had a Stone Age economy.' He thought the site might
have been an Early Iron Age pottery. The Golf Club committee
agreed to re-site their green so that the area remains intact, and
it will be possible to excavate further in the immediate vicinity.
It stands just above Rue des Mielles, the road running from the
Nunnery to Whitegates.

The Roman Find of 1972

Nearly opposite Les Huguettes, on the fringe of Longy
Common, stands the house known as 'The Kennels', where
what was thought to have been the remains of a Roman building were found in 1972. Bones and pottery were discovered, and
Kenneth Wilson, investigating, came across a fragment of
smooth red Roman samian ware, which he thought was part
of a grinding pot not made before about AD 150. He also found
pottery with the same sort of decoration as the pots found at
Les Huguettes. 'The radio-carbon date of this site is 490 BC',
he stated, 'Thus we had dated the layers immediately above
and below the wall found at "The Kennels".' He thought the
building Roman in character, but did not think it had had a
very long life and believed that most of its stone was used elsewhere.

THE OLD TOWN

Louisa Lane Clarke in *The Island of Alderney* (the first guide to
the island, published in 1851) stated that Alderney's 'old town'
was in the north-east (the Longy region) and that it was
destroyed 'ages ago by the judgment of God, after a Spanish
ship had been wrecked on the coast and its crew most barbar-
ously murdered'. Another belief was that sand blown by pre-
vailing north-east winds gradually covered the town, causing
its inhabitants to retreat to St. Anne, where another town was
established.

Modern historians are inclined to treat these statements as
folklore, though recent archaeological excavations seem to point
to an excess of sand as a contributory factor to the decay of
'Longis', as it is still sometimes spelt; but certainly there was no
abrupt exodus to St Anne. Longy in prehistoric times must
have been a settlement as well as a necropolis, probably lasting
from the Megalithic Period (c 3000–1000 BC) to the Bronze Age,

judging by the discoveries, including about 2,000 bronze implements, on Longy Common in 1832.

This is probably Alderney's most ancient building; its name was supposedly given by some military wag of long ago who likened its solitude to that of a convent. The name has stuck and its old one, 'Les Murs de Bas', is as forgotten as 'Les Murs de Haut' for Essex Castle.

Lt-Col C. S. Durtnell, who excavated at the Nunnery in 1929 when stationed in Alderney with the Queen's Own Royal West Kent Regiment, expressed the belief that the Nunnery was a fort at Alderney's most vital point. Here at Longy was the island's only port, until Braye supplanted it in 1736, and even today the remains of its pier can be identified at low water.

Writing on the subject in the *Guernsey Evening Press* in 1966, Lt-Col Durtnell was of the opinion that the Nunnery was built on top of a Roman kitchen midden, which he had examined in 1929. He stated that obviously no Roman would have built anything on top of his own midden, but 1,000 years later, in 1436, the ground would have settled and there would have been no disadvantage in building. 'In fact', he wrote, 'the builder of Les Murs de Bas must have been quite unaware of the fact that a midden existed underneath the foundations'. He concluded: 'Sandwiched one on top of another are the remains of six civilisations: Medieval; Norman and Viking; Saxon; Roman; Celtic; Pre-Celtic (? Iberian).' In 1972 the material he had found and presented to the Guille-Allès Library's museum in Guernsey was returned to Alderney by the museum authorities and is now displayed in the museum of the Alderney Society.

The belief that the Nunnery was substantially a Roman fort is disputed by D. E. Johnston, staff tutor in archaeology at Southampton University, in a contribution to the Alderney

Society's *Bulletin* for December 1971. He holds that no written evidence goes farther back than about 1540, 'and the curtain walls and bastions could well be medieval. Only excavation can show whether there was once a gateway of Roman type beside or beneath the present 18th century one'.

Mrs Lane Clarke boldly styled the Nunnery 'the ancient *Castrum Longini,* or Château de Longis', evidently raised with the materials obtained from the 'Roman town, situated a short distance from its walls'. It was restored for use as a barracks in 1793. She stated that it was the residence of the Chamberlain family and the early governors of the island, 'but the house was probably pulled down in 1793'.

Sir Thomas Kendrick, in *The Archaeology of the Channel Islands* (Volume 1), mentioned 'the massive fragments of Roman tile embedded in the core of the walls and the herringbone work still visible'. While these all favoured a Roman date, he cautiously added that none of these features was a decisive criterion, since herringbone work could have been Norman as easily as Roman.

In 1889 Baron von Hugel and Dr F. P. Nichols excavated what became known as the 'Longy Refuse Pit', north-east of the Nunnery. Kendrick wrote that 'the place was a rubbish dump of the Roman period'. It yielded, among other things, fragments of nearly 100 pottery vessels, a glass bead, chips of glass, bricks and tiles, as well as a coin of Commodus.

Burials near the former Coastguards' buildings (between the Nunnery and Whitegates) excavated in 1905 yielded fragments of a grey Gallo-Roman ware and a Roman coin. Skeletons wearing bronze collars were also found there. Whether these discoveries, together with the 'finds' at 'The Kennels' and the burial on Raz Island, constitute evidence that Alderney was occupied by the Romans is still not clearly established.

5   THE MIDDLE AGES AND LATER

FROM the Roman era to the Norman Channel Islands history is practically silent. Written records do not exist and one is forced to rely on legends (of doubtful worth), folk memories (slightly more reliable) and personal deductions in order to bridge the gap between the two civilisations. No one can say with certainty when Christianity was introduced into the islands, when organised government began, or when, indeed, they were colonised.

The belief that Christianity was brought to Alderney by St Vignalis has been discredited by such historians as A. H. Ewen, who believes he went to Herm rather than Alderney. This authority, writing in the official guide to the Island, pointed out the sites of various chapels, including one at Mannez; another near St Esquerre Bay (close to Houmet Herbé); one near the Strangers' Cemetery of St Michel, on the Longy road and adjoining Les Rochers; and perhaps one near Rose Farm (facing the airport), close to a pagan holy place at l'Essource (the source of the island's largest stream).

These chapels were probably built on or near pagan sites so as to counteract their ancient influence on the islanders. Such buildings would have been set up by those who brought Christianity to the Islands, very likely in the sixth century, and especially by Sts Sampson and Magloire, though there is no evidence that either visited Alderney. Perhaps the Celts went there, though they left no positive evidence of this in the island placenames, unlike the Normans, whose names, ending in 'ey' or 'y', denote large islands (Jersey, Guernsey, and Alderney or

49

Aurigny), as well as those ending in 'ou' (Brecqhou, Jethou and Burhou), meaning small islands. Their name for large rocks was *etacs*, while a tide race was a *raz*.

Clearly, then, the Norsemen settled in Alderney as the forerunners of those Normans who, conquering that part of northern France they called Normandy, annexed the Channel Islands to what had become their duchy in 911, under Rollo, their first duke.

ROYAL AND RELIGIOUS RULE

In a charter of 1042 Alderney was granted by Duke William (later the Conqueror) to the Abbot of Mont St Michel in compensation for the half of Guernsey of which he had deprived the religious house. In 1057 William transferred the grant of Alderney to Geoffrey de Mowbray, Bishop of Coutances, and gave the Abbey of Mont St Michel land at Noirmont in Jersey. A charter of 1054 stated that the church in Alderney was granted to the church of St Marie in Cherbourg, while another charter, dated 1134, named the Alderney church as being dedicated to Ste Marie.

For the next three centuries the Bishops of Coutances retained much of Alderney, a document of 1236 revealing that the then bishop owned half of it. The other half remained in ducal hands, although by then the duke had become king of England, and thus Alderney, with the other islands, became linked with, but not subject to, the English Crown, for the islands continued to be Norman 'overseas dominions'. When King John lost Normandy in 1204, the islands remained associated with England, while France changed from a friendly relative into a traditional foe, remaining one for centuries. Despite this state of affairs, the Bishop of Coutances remained the spiritual head of the islands until the Reformation.

Unlike the other Channel Islands, Alderney was not under feudal rule. There was no *seigneur* (overlord), resident or absent,

whereas in the larger islands there were many. In Sark, more-
over, Elizabeth I created a *fief* (territory) held by the Crown
and leased to individuals in return for payment or services rendered) yet Alderney, of comparable size, was administered differently. There appears to be no known reason for this peculiarity.

A document drawn up about 1236, the *Status Insulae de
Aurineo*, set out the rights of the Crown and the bishop in the
island and, in so doing, afforded an insight into its appearance
and way of life at the time. The king had his Court there, comprising six jurats (judges) and a *prevôt* (sheriff), who enforced
the laws. Wheat growing and sheep rearing were the farmers'
main activities. There was a watermill owned by the bishop,
the king had a windmill (one of the earliest recorded references
to windmills in western Europe, according to Mr Ewen),
fishing rights were leased out and there was a priest in residence.
In 1274 Alderney's Crown revenues totalled £60 9s 2d.

The Assize Roll of 1309 records names such as Duplain,
Houguez, Le Vallée and Pezet, some of which still survive in
Alderney. By 1320 the jurats had increased to twelve and the
Chairman of the Court was the senior jurat. Later the number
was reduced to seven, as it is today.

MEDIEVAL AGRICULTURE

The open-field system is of great antiquity, and Mr Ewen, who
has made an intensive study of the subject, has stated that 'it
became widespread in north-western Europe in the period
following the collapse of the Roman Empire in the fifth century
A.D.' Precisely when it was introduced to Alderney is, of course,
unknown. It was located in the area known as La Blaye, a word
doubtless related to *blé*, meaning a cultivated cereal crop. La
Blaye adjoins St Anne and is south of it. From Le Giffoine in
the west it extends over high ground to La Haize in the east,
with the southern limit formed by a boundary running almost
to the cliff edge.

It was divided into *riages*—rectangular areas which, in turn, were subdivided into strips. These were the property of the *riagers*. The strips still exist, though they are much smaller than they were originally, owing to subdivision among heirs of properties. Beyond the Blaye was the non-arable land—the commons—where cattle and sheep were grazed and where furze and bracken might be cut. Earthen banks kept cattle out of the Blaye while it was under cultivation, but at other times the animals were free to graze there, thus manuring the soil before the spring sowing. Additional fertiliser came in the form of *vraic* (seaweed), which was carted from the shore and spread over the land.

The *Douzaine* (parish council), which no longer exists in Alderney, though it does in the other islands, administered the Blaye. Its members inspected it twice yearly, settling disputes as well as ensuring that gates and hedges were in good condition. An interesting drawing of the Blaye is included in *The History of Alderney* (1810) by E. A. W. Martin. It shows corn being grown in strips, with St Anne and its old church tower in the background. The whole area covered some 450 acres, including La Petite Blaye, and the *vaindifs* (extra strips to allow the plough to turn at the end of a furrow).

NUCLEUS OF ST ANNE

It would seem that at one period the Bishop of Coutances lost his rights in Alderney, for the 1361 Treaty of Brétigny between England and France restored them to him, although more in theory than practice, apparently. Still, he continued to appoint his priests until 1568. Edward Portman, of Salisbury, leased Alderney from the Crown in 1376, and during the following century many Guernsey folk settled there, bearing such names as Ollivier, Le Cocq and Simon, names still surviving. Others, including Herivel, Gaudion and Bott, came from Normandy in

Page 53 (*above*) The Iron Age site, recently discovered near Longy Bay. Beyond is Essex
Hill and the valley known as Essex Glen. Adjoining the site is the golf course

Page 54 (above) The Nunnery, once styled *Le Chateau de Longis*, stands beside the shores of Longy Bay. Above is Essex Castle, the more modern barracks on the left contrasting with the medieval walls and early nineteenth-century watch tower. Roman work in the Nunnery's masonry probably makes it Alderney's most ancient building; (*below*) From the south-east cliffs Longy Bay is well seen. Offshore is Raz Island, with its Victorian fort, and beyond is Fort Houmet Herbé of the same period

the fifteenth century and their descendants are still to be met with in Alderney.

This immigration caused the original township of St Anne to expand. It still retained its rural atmosphere, with thatched farmhouses and outbuildings lining the streets, but the original settlement in and around the present Marais Square, near the Blaye (where there is water), was extended to Le Bourgage, a region east of the Marais and running parallel with the High Street, where there were wells. The narrow lanes, known as *venelles*, provided rights of way to this water supply for those living in the Bourgage, as well as affording easy access to and from the Blaye. The High Street itself did not become important until the eighteenth century, although it must always have been a thoroughfare of sorts, since it was the way leading to Longy, the ancient port of Alderney.

Rue des Vaches, now Little Street, is an ancient way connecting the arable with the township and, as its name suggested, along it passed cattle from the Blaye on their way to water perhaps or to their stalls. Down this hill flowed the stream feeding the Marais wells, before continuing its course to what is now Royal Connaught Square and then flowing down Le Pré and the valley to the sea at Saline Bay. Another old region is Le Huret, between the High Street and Marais Square, where Alderney's first Courts, the king's and the bishop's, were held in the open. It was not until 1770 that meetings took place in a courthouse, a rude building standing on the site of the tennis courts at the rear of Ralph Duplain's Victoria Street premises.

Gradually the town expanded, incorporating Rue des Sablons (Sandy Lane), later to be called Rue Grosnez and, later still, Victoria Street. Perhaps the lanes between Le Marais and St Anne's Square (later renamed Royal Connaught Square) came into use during the seventeenth and eighteenth centuries, while the lower High Street (between the top of Victoria Street and Le Huret) may date from about the same period.

D

The medieval windmill belonging to the king stood on the Blaye, very near Burland's Farm, and is shown on the Chevalier de Beaurain's map of 1757. The bishop's watermill was probably the forerunner of that standing at the foot of Le Petit Val, inland from Platte Saline. It was fed by a stream running down Le Val de la Bonne Terre from Pont Martin (on La Petite Blaye) to the sea. The 'Big' and 'Little' Blayes probably came into being when the whole of the original Blaye was divided by the building of Le Grand Val road, running west from St Anne.

LONGY

It would seem that Longy dwindled when the agriculture of the Blaye became more important than its fishing, but it retained a measure of significance until about the end of the eighteenth century. A 1757 map, the work of the Chevalier de Beaurain, Geographer in Ordinary to the King of France, marks several buildings along the shore, including what must have been the *Castrum Longini* (also marked on Leland's map of Alderney, 1550). De Beaurain styles it *Le Fort* and also marks *Le Château* (the present Essex Castle), as well as a jetty, traces of which can still be seen at low water. His map shows no buildings at Braye, nor a pier. Longy is linked with St Anne by a well defined road, which forks near the approaches to the shore. One arm leads to the port and the other to Essex Castle. A much earlier map of 1540 in the possession of the Bodleian Library, Oxford, marks *Castrum Longini* on an island named 'Alderney', perhaps the earliest reference to the modern alternative of 'Aurigny'.

In the reign of Henry VIII, because of war with France, it was resolved to build a powerful fort which would dominate the approaches to Longy. Not only would it defend the island but it would enable Longy Bay to become a place of safety from French privateers and pirates. Work on this enterprise began in 1546 and over 200 men were sent from England to build what later became known as Les Murs de Haut and, later still, Essex Castle. The labour force was directed by Robert Turberville, who rejoiced in the title of Marshal of Alderney. The work languished after the deaths of Henry and the Protector Somerset, and ceased in 1554.

Today, the northern and western sides of the outer wall are those built in Tudor times. The citadel, standing in the centre, comprised four donjons, but this powerful building was destroyed, with the eastern and southern walls, when the Victorian Fort Essex was built.

During the Tudor building period Alderney had a garrison, which may have been quartered at Les Murs de Bas (the Nunnery), but when operations ceased, the troops were recalled. As a result, Alderney was invaded by a French expedition from Cherbourg in 1558, and cattle and other booty were carried off. The leader, Captain Malesart, resolved to repeat the exercise, but the English fleet appeared in the offing and he withdrew. He returned again but encountered a force sent from Guernsey, which captured him and despatched him to the Tower of London.

THE CHAMBERLAINS

Guernsey's Governor at that time was Sir Leonard Chamberlain, and before his death he obtained a grant of the governorship of Alderney for his son George, who, like the rest of the family, was an ardent Catholic. In 1570, because of this, George was obliged to flee to Flanders, but in 1584 his brother John was granted a patent, receiving Alderney on payment of £20 and an annual rent of £13 6s 8d. John Chamberlain is believed to have turned the Nunnery (to use its subsequent soubriquet) into his own residence. His differences with the people of Alderney were many and resulted in a series of ordinances setting out his rights and those of the inhabitants.

In 1591 Chamberlain sold his rights to the Earl of Essex for

£1,000, but there is no evidence that Essex ever went to the island, despite the fact that Fort Essex was named after him. He was executed for high treason eight years later, but just before this he had leased Alderney to William Chamberlain, John's brother; but William, too, ran into trouble with the islanders, and as a result the differences were examined by Royal Commissioners in 1607, when most of the people's wishes were granted.

The Chamberlains remained Governors of Alderney until 1643, just after the start of the Civil War. The castle on the hill fell into decay, but the Nunnery apparently remained in good condition until 1793, when the residence was demolished to make way for barracks. Its ancient walls, however, were spared.

SIXTEENTH-CENTURY REFUGEES

Among Alderney's surnames are those of Audoire (or Odoire), Batiste, Renier and, until recently, Barbenson (originally the Flemish Brabançon). Their original holders fled to Alderney as Protestant refugees from France in the reign of Edward VI. They were welcome, for they spoke a language understood in Alderney, and shared the post-Reformation Calvinism of the islanders. With services held in French instead of Latin, English ministers were useless there, so that by 1576 the Presbyterian system of the French Reformed Churches was adopted in the Channel Islands, thanks to the French pastors who fled there. This adoption met with the approval of the Council of Elizabeth I. Guernsey and Alderney maintained this system until the reign of Charles II.

SEVENTEENTH CENTURY

During the Civil War, Alderney, like Guernsey, was held by Parliamentary forces. Captain Nicholas Ling was appointed Lieutenant-Governor and commander of the island Militia in

1657. This force had been in existence for centuries, like those of the other islands, but Ling seems to have been its first recorded commander. He retained his offices until his death at the age of eighty in 1679. At the Restoration his tenure of office had been renewed, perhaps because his wife had been an Andros, a Royalist family in Guernsey.

Ling built his Government House in St Anne's Square, the building known today as the Island Hall. By this time Alderney was no longer under the authority of the Governor of Guernsey (a Civil War measure) and Charles II granted it to the Royalist Jerseyman Sir George Carteret. He and his Lieutenant-Governor, Ling, died within a few days of each other, whereupon Lady Carteret sold the patent of the island to Sir Edmund Andros, of Guernsey, who bought from Ling's heirs 'the house, garden, enclosure, meadow and furze breake', wherein to accommodate his own Lieutenant-Governor.

THE LE MESURIER REGIME

Thomas Le Mesurier, a kinsman of Andros, filled this office and his son, John, married Anne Andros (niece of Sir Edmund) in 1704. Dying childless in 1713, Sir Edmund bequeathed the Alderney patent and house first to his nephew George, and then to George's sister Anne, wife of John Le Mesurier. The people of Alderney, however, resented a Guernsey family as their Governors and relations became strained.

Anne died in 1729 and Henry, her eldest son, succeeded her. Seven years later he built what is now the old jetty at Braye, thus raising it to the status of a harbour, the idea being to make a fortune in smuggling, as others were doing in the islands. He married his cousin Marie Dobrée, also of Guernsey, in 1742, but she found life in Alderney so difficult that her husband transferred the patent to John, his younger brother, in exchange for property in Guernsey.

John appears to have found it equally hard to live with

Alderney folk, but privateering altered matters. During the continuous French wars he fitted out ships manned by islanders; Braye harbour accommodated them, warehouses for 'acquired' merchandise were built there and fortunes were made. The Le Mesuriers had become popular almost overnight.

As hereditary Governors they commanded the Militia from 1684 until 1803, apart from two intervals during which local officers assumed control. Until 1781 only the officers wore uniform, and they supplied their own. The strength of the Militia seems to have been about 200 during the period under review, but by 1850 it had been reduced to 150, of whom 100 were infantry and the remainder artillerymen. In common with the other Channel Islands' Militias, that of Alderney was made Royal in 1831 to commemorate the fiftieth anniversary of the victorious Battle of Jersey, in which the Militia helped to defeat a French invasion.

Although the infantry was disbanded in 1882, the Alderney Militia Artillery continued, and an arsenal for this force was built in Ollivier Street in 1882. It was demolished about 70 years later. The Militia was abolished in 1929, following the decision of the British Government to cease paying the cost of the Channel Islands Militias, apart from that of Sark, which had come to an end in 1880. The Militias of Jersey and Guernsey continued, though greatly reduced in strength, until 1940.

EARLY DEFENCES

The fortifications of Longy apart, Alderney was not devoid of defences before the building of the Victorian forts. William Chamberlain, reporting to the English Council in 1627, was apprehensive because Alderney's defences had become so weak. In the past, their cost had been met by the 'Farmer' (Crown Tenant) of the island and the inhabitants, but since there were only 800 of them, and they were poor, the Council was asked to supply Alderney with powder and munitions.

60

In 1739 John Bastide drew a survey map of the island that lists sites where it was proposed to build batteries mounting altogether twelve 12pdrs and fifteen 6pdrs—at Longy, Corblets, Château à l'Etoc, Roselle Point, Braye Bay and harbour, Crabby, Saline Bay and near Rocque Tourgis, overlooking Clonque. Bastide's recommendations were adopted about a century later, but on a far larger scale. Yet clearly earlier fortifications were in existence, some of them armed with cannon from the wrecked frigate HMS *Amethyst*, for they were mounted in a 'new' battery in 1795, according to Mrs Lane Clarke's Alderney guidebook. The guns serving as bollards on Braye's old pier and at the New Harbour are supposed to have come from this warship.

In Martin's *History of Alderney* (1810), it was recorded that the States (the local governing body) built and kept in repair the batteries and guard-houses, although 'about three years ago Government was pleased to grant the sum of £20,000 for the purpose of erecting barracks for 800 men and a further sum for a hospital'. Martin suggested building a 'Place of Arms' on Mont Touraille, and nearly half a century later this proposal was adopted.

A map drawn in 1831 revealed artillery barracks at Butes (the site of an ancient archery ground), and on its northern slopes was a battery. Another was marked in the centre of Braye Bay—'St Anne's battery'—and perhaps this was the same as the one known as 'Nine Guns battery', traces of which could be seen until a few years ago emerging from a great heap of spoil from the adjacent Battery quarry.

There were batteries in the centre of Saye Bay and at Château à l'Etoc, while at Quèsnard was a guard-house. At Longy a battery and barracks (at the Nunnery) were shown on the map and a guard-house stood near the Hanging Rock, close to Essex Castle. Two more were on the cliffs of Plat Côtil and Trois Vaux.

In 1833, James Wyld, Geographer to the King, published

61

his excellent map of Alderney, and as well as marking these fortifications he showed barracks between Essex Castle and the Nunnery. Possibly this is the origin of the name 'Barrack-Master's lane' (otherwise Essex Glen). Wyld also mentioned a 'King's battery' south of Corblets. Another, seaward of 'Touraille on the Mount', might have been at Roselle. On the Houmet de Clonque stood a battery, with barracks nearby. A little later, batteries were installed on the sites of Forts Doyle, Platte Saline and Tourgis and there was a signal station on Mount Touraille. There were gun positions near the Nunnery and below 'Fort Hill', south-west of that building.

Further defences were the beacons set up around the coast ready to be fired at the enemy's approach. The best-known was that at La Giffoine, on the western cliffs, but there was another on 'Fort Hill' and two more down at Longy, according to the 1824 map drawn by Captain M. White, RN.

Telegraph Tower

This stands on the western cliffs and it was used in the early nineteenth century to communicate with Guernsey and Jersey, through the medium of Sark (where a repeating telegraph was established) in the event of suspicious ships approaching the islands. Martin (1810) stated that the tower had been 'lately erected', without being more specific, though the telegraph was a new construction and the work of 'the ingenious Mr Mulgrave'. The building later became ruinous, but was subsequently restored and today serves as a seamark.

THE EIGHTEENTH-CENTURY TOWN

According to Mr Ewen, the town of Alderney became known as St Anne early in the eighteenth century, while Simon Masson was its minister. It had been known as St Mary since Norman times, and possibly the 'new' name was adopted because the old church contained a chapel dedicated St Anne.

The town, at this time, did not present a very impressive picture. Near the church, which was small and of indifferent architecture, stood Government House. The other buildings were humble, with thatched roofs and a distinctly rural appearance. The scene was somewhat enhanced by the church's clock tower, added in 1767 and the building itself was slightly improved in 1790 by the addition of a gallery at the west end. The clock tower, happily, still stands, though the rest of the church has been demolished.

Les Mouriaux House, opposite what is now the Island Hall and a short distance beyond the Royal Connaught Hotel, was completed in 1779, and is one of Alderney's most elegant residences. It was built by Peter Le Mesurier, and on his father's death in 1793 it became the Le Mesurier family home. Government House ceased to be a residence, though its martial use remained until the end of the Napoleonic Wars. Les Mouriaux House became the home of the first President of the States of Alderney, the late Captain S. P. Herivel, RNVR, after World War II.

Early records make no direct reference to the States and it is quite impossible to say when this body was established. Like the States of Jersey and Guernsey, it must have evolved from the Court and *Douzaine*. The word 'States' is said to have referred to the three estates of the realm.

Originally, the Court had met in the open air at Le Huret, but in 1770 it moved to slightly more comfortable quarters in a building off what is now Victoria Street. Here the jurats met every Saturday and dealt with civil causes only, according to the historian Martin, for criminal cases had to be referred to the Royal Court of Guernsey. Hence the absence of a gaol in Alderney until one was built, alongside the present Court House, in 1850, by which time the Court of Alderney's powers had broadened in scope.

In 1790, Governor John Le Mesurier generously founded the Town School, and the fact is recorded in French in an inscription over the entrance to what is now the Alderney Society's museum. In her book Mrs Lane Clarke revealed that the wife of the last of the Governors established the Mouriaux School as a thank-offering for her first-born 'and with her own hand wrote out the rules and lessons still used there'.

END OF AN ERA

A curious incident occurred in 1792, when 124 French priests took refuge in the island, where they remained until 1793, during the height of the French Revolution. They came from Jobourg, across the Race, and landed at St Esquerre Bay, near Longy. They found accommodation in St Anne and were hospitably received by Peter Le Mesurier, who was then acting Governor for his infirm father, who died in 1793. Peter Le Mesurier did more than welcome them, he allowed them to worship as they wished at a time when religious tolerance was at a particularly low ebb.

However, they were asked to be discreet and not disturb the islanders, who were, for the most part, Wesleyans, thanks to the missionary work of Dr Adam Clarke. Pierre de Solier, a Frenchman, was then the Anglican minister, and he, too, was tolerant of the refugees, like the majority of the community.

Peter Le Mesurier succeeded his father in 1793, but in 1795, while he was superintending the mounting of a gun in a new battery, it fell, seriously injuring him. He died in 1803 and his son succeeded him. He was General John Le Mesurier, who was neither popular with the Guernsey military authorities nor the local Court. He disputed the right of the former to interfere with the defence of Alderney (claiming this to be his prerogative) and fell foul of the latter because he wished to be known as *Seigneur* of Alderney. A majority of jurats refused to acknowledge this title, so Le Mesurier was forced to drop it.

Smuggling had been an important source of revenue in the Channel Islands, especially in Alderney, with its relative proximity to England and France, but in 1808 the British Government took such positive steps to suppress it that Alderney was seriously affected. The end of the wars with France, following Napoleon's defeat at Waterloo in 1815, also meant the end of privateering and the withdrawal of the garrison. A local bank failed and, in general, there was widespread poverty in the island. None of this pleased John Le Mesurier who, in 1825, surrendered his patent to the Treasury for an annual payment of £700 during the remainder of his lease. The family left Alderney and their departure added further distress to the island.

This was somewhat alleviated in 1830, when it was agreed that the Alderney common lands should be divided between the islanders proper. The commons were divided into good land and poor land, and each of the fifty-two families received a good and a bad lot. By then the population numbered about 1,000, but even with this land in their hands the future did not appear at all inviting.

VICTORIAN DEFENCES

IT is curious that the inception of tourism in Sark and Alderney began in the 1840s and that it was primarily due not to scenic attractions but to silver mining in Sark, which captured the imagination of the tourist to Guernsey, who went there by sailing cutter to inspect the workings, and to the massive fortifications in Alderney, which again drew tourists from Guernsey by sailing vessel. The tourist influx also resulted in the establishment of a steam packet service, which induced others to visit Alderney from Guernsey. Today it must be admitted that the Victorian forts of Alderney still add character to the landscape.

THE FRENCH MENACE

In 1842 the British Government became perturbed at the fortifications the French were building at Cherbourg, which included a great naval harbour and several forts to protect it. Cherbourg is only about 25 miles from Alderney, and as a counter move Britain resolved to make Alderney 'the Gibraltar of the Channel'. This decision took some time to implement and it was not until 1847 that work began on what was described as a 'harbour of refuge' in Braye Bay. The intention was that this should form a naval station and would be the counterpart of Portland harbour, on the other side of the English Channel.

The original plan was to construct a harbour resembling that of Cherbourg, enclosed by two arms—the one existing today, known as 'the breakwater', and another which would have extended from Roselle Point on the north coast towards its fellow. These arms would have enclosed 67 acres, but in 1858 it was resolved to enclose 150 acres by building an arm seaward from Château à l'Etoc. In fact, though most of the western arm was completed, the eastern section was never begun.

The *Illustrated London News* of 20 February 1847 reported on what was described as 'laying the foundation stone' of the Alderney harbour works, though this just involved tipping a great boulder into the sea, the first of many to form a breakwater. This action was accompanied by a salute of guns, fired by the Royal Alderney Militia Artillery, while the infantry of the Militia watched. Officials from Alderney and Guernsey, together with many other folk, walked in procession from St Anne to Braye and, the ceremony over, all enjoyed feasting and fireworks that evening.

It was the existence of the naval harbour that led to the forts' construction and lent Alderney the title of the 'Key of the Channel'. The harbour's construction, together with the building of its defences, attracted thousands of workmen to the island and additional living accommodation had to be built for them. Walker & Burgess were the breakwater's engineers and Jackson & Bean the contractors. In order to accommodate vessels engaged on construction work, the New Harbour was built at Braye, not far from Crabby. Today it is used by local fishermen and yachtsmen, but in the last century it accommodated the steam tugs *Bolton* and *Watt*, used by Jackson & Bean for towing barges laden with stone and other equipment to the works, and by the paddle steamer *Queen of the Isles*, built for the contractors and in inter-insular service from 1853 to 1872. She was the first steam packet to carry passengers between Guernsey and Alderney.

Work on the breakwater was often hindered by the sea's action. Westerly gales sometimes wrought havoc on the mole and it was said (probably justly) that the breakwater was built contrary to the advice of Alderney pilots, who declared it would never withstand the onslaught of westerly gales. It has done so, but only after severe damage has had to be made good, frequently and at great expense, since work began in the 1840s.

The Alderney developments captured the interest of the British public (even if Channel Islanders were less concerned), and it is not surprising that Queen Victoria herself paid the island a visit. She arrived with the Prince Consort aboard the royal yacht *Victoria and Albert* on 9 August 1854, and during a brief stay inspected Fort Albert and the breakwater. The Guernsey artist Paul Naftel was commissioned by Her Majesty to paint the scene, and his fine work shows the fleet in the road-stead, the royal couple landing at the breakwater slipway, and the Royal Alderney Militia lining the route, with hundreds of spectators nearby.

The Alderney breakwater and forts attracted the Queen and Prince Albert again in 1857 when, on the evening of 8 August, they arrived with other members of the royal family on the royal yacht, which was escorted by a fleet of warships. The visit was unexpected and hurried arrangements had to be made by the authorities to receive the distinguished visitors. The Queen and Prince Albert landed at the breakwater the following morning and travelled throughout the island, by railway car and carriage.

The railway car must have been specially adapted for the carriage of royalty, for normally the train was used in the haulage of stone from Mannez quarry to the breakwater, with stops along the line. The track still exists and is occasionally used by a string of trucks, hauled by the diesel locomotive *Molly II*, laden with huge blocks of stone which are deposited on the seaward side of the breakwater to break the force of the waves. This operation has continued ever since work on the mole began, probably on 14 January 1847, when the railway

68

began running with fourteen trucks drawn by an engine from Mannez to Grosnez Point, as the *Jersey and Guernsey News* reported.

The construction of the breakwater and forts went on for about 30 years, and by 1872 the breakwater had been extended 1,600yd. In that year a Select Committee of the House of Lords engaged on examining the Alderney defences had doubts about the project. It reported: 'If such a work were now for the first time proposed, with the experience of its difficulties and results, the most eager advocate for national defences would probably hesitate in recommending its commencement.' A similar view had been expressed much earlier by Captain Christopher Claxton, RN, when he gave evidence before the Select Committee on Harbours of Refuge in 1858. He said: '. . . there is not room for seven ships in the whole space they have provided, except they bump against each other. They have made works which require 3,000 men to man the guns, and they have not 700 there.' When asked if a couple of sail putting out of Cherbourg could take Alderney before men could possibly be sent to defend it, he answered: 'Certainly, if there were no more there than there are now.'

The 1872 Report of the Select Committee speculated on the wisdom of maintaining the fortified harbour in whole or in part, and whether it should be destroyed or 'obliterated'. For by this time the French menace, if it had ever seriously existed, had passed, and the British Government was having second thoughts about Alderney and the immense amount of money it had already cost the British taxpayer.

The experts who had visited the breakwater in 1870 had suggested that it be maintained for a limited period. It had cost over £1¼ million to construct and maintain over the first 25 years of its existence. The forts and batteries had cost £260,000 to build, and the experts suggested that two forts would suffice—doubtless Forts Albert and Tourgis—but none, in fact, were demolished.

The Select Committee recommended the 'temporary' maintenance of the breakwater, and this has gone on ever since! Part of it has been allowed to disintegrate and today only about half the original mole is maintained. A submerged section is given a wide berth by shipping entering or leaving Braye.

Today the breakwater is maintained by the Ministry of the Environment, and formerly it was the responsibility of the Admiralty. It costs Alderney nothing, yet its value to the island is incalculable, for without it the jetty (built at the turn of the century for merchant shipping to use) and to some extent the New Harbour would be exposed to the full fury of westerly gales, while the roadstead would be almost useless.

On only one occasion did the so-called naval harbour justify its existence—in 1901, when the Channel Fleet used the anchorage during summer manoeuvres. The forts, on the other hand, were used to a limited degree by the War Department and the island Militia.

THE CHAIN OF FORTS

The Victorian forts of Alderney were designed by Captain (later Lt-Gen Sir F. W. D.) Jervois, who allowed his artistic taste to get the better of strictly military needs at times. In consequence, one finds medieval embellishments such as arrow slits and machicolations, battlements and moats, more in keeping with the defences of Elizabeth I than those of Victoria. The finest quality dressed stone was used and care was taken to use stone of different shades, thus vastly improving the appearance of the forts. The setting, of course, further enhanced their appearance. In fact, the military architect had the time of his life in Alderney!

Today some of the forts are the property of the States of Alderney, others are in private hands, and a few are in ruins. At the time of writing, the States are undecided as to the future use of Forts Albert and Tourgis. Essex Castle and Château à l'Etoc have been converted into residential flats, Fort Clonque

Page 71 (*above*) Fort Clonque as the gannets see it. The Victorian defence work is linked with Alderney by a causeway, covered at high water. The nineteenth-century fort was modernised by the Germans in World War II and is now let as holiday flats; (*below*) Saye Bay, with Fort Albert beyond. Once used by troops, it is now the haunt of the holiday-maker. Though never crowded, its sands are perhaps the most attractive in the island

Page 72 Victoria Street is Alderney's principal shopping centre. Its paving stones add to its attraction, while the small shops on either side enhance its charm. Its old name, *Rue Grosnez*, was changed in honour of the royal visit of 1854

provides tourist accommodation, Fort Grosnez is used for breakwater maintenance and Fort Platte Saline forms part of a company's premises. Fort Corblets is one of the island's most imposing private residences and the fort on Raz Island is also in private hands.

Fort Albert

This was easily the most important of the Alderney forts, for its position commanded the approaches to the naval harbour. Mont Touraille, as the hill on which it stands is called, used to be a signal station, and it is only surprising that the position was not fortified in Napoleonic times, if not earlier. Viewed from the sea, Fort Albert's stone merges into the landscape so well that its existence is hardly noticeable, and the other forts, built of native granite, are similarly well camouflaged. Fort Albert does not wear such an impressive appearance as Fort Tourgis, though seen from the roadstead it looms above shipping in the grand manner of a medieval castle. It was named after the Prince Consort, possibly in remembrance, after his death.

Fort Albert was the kingpin of the chain of defences, which were built from Clonque on the west to Quèsnard on the east and Essex in the south. The forts, started about 1847, were intended to defend Alderney at its weakest points, so none were constructed on the steep southern cliffs. Between Clonque and Essex thirteen forts were built along a low-lying strategically weak coastline. Some occupied the site of earlier defences. In this respect, it is interesting to observe that the Germans, when they fortified the island during the 1940–45 occupation, also built their defences at positions which, for centuries past, had been regarded as militarily important.

Like all its fellows, Fort Albert was primarily intended to be an artillery position, but in practice, and following the disuse of most of the other forts, it became the military headquarters of the island, housing the infantry garrison until that was withdrawn in 1929. Even after that year, troops stationed in Guern-

E

sey periodically visited Alderney for exercises, and were accommodated at Fort Albert. Its guns were also used for training the Royal Alderney Militia Artillery.

The fort occupies a considerable area of Mont Touraille, and has a spacious parade ground and much barrack accommodation within its walls. The position of its guns can still be seen. Fort Albert suffered grievously from the Germans, and a long period of neglect after World War II did not improve its condition.

An Army list of Alderney defences in 1886 reveals that sixteen relatively heavy guns were mounted in the fort. Later, this number was reduced and some modern 6in guns were installed. Further guns were mounted at Roselle battery, on a headland below the fort and commanding the harbour, and here, during World War I, searchlights were installed to cover the approaches of Braye.

Fort Tourgis

Like many other Alderney forts, Fort Tourgis bears a local and family name. It stands on a site once known as Rocque Tourgis on a low headland separating the bays of Clonque and Saline. Beside the main building is a stone cone, whitewashed, serving as a seamark and replacing the stone which once assisted mariners in their navigation of the Swinge, that hazardous channel between Alderney and Burhou.

Like the majority of its fellows, Fort Tourgis was built largely of stone quarried in the vicinity, a procedure that not only saved labour but ensured camouflage. Some material was brought either by train or horse-drawn vehicles from quarry to site. An enormous amount of stone must have been used, for Fort Tourgis is second in size only to Fort Albert.

Visiting troops on exercises in Alderney today usually make use of Fort Tourgis as a barracks. In the past it could house a considerable number of troops, and reasonably good roads permitted their quick transport to any other part of the coast

threatened with invasion. In 1886 Fort Tourgis mounted sixteen heavy guns (chiefly 8in), this artillery complementing the
cannon of the smaller Forts Platte Saline and Doyle nearby.

The architecture of Fort Tourgis is, many will agree, superior
to that of Fort Albert. Although of considerable extent, there
is nothing ungainly about the fort, and the silver-grey quality
of its stone is handsome. Despite its good looks, it has enjoyed
little history, and apart from its use by the Alderney Militia
(and, of course, by the Germans), Fort Tourgis never attained
the importance of Fort Albert—though while the latter was
being built, the smaller fort housed Alderney's garrison.

Forts Platte Saline and Doyle

These were relatively minor defences, more batteries than
forts, standing near Fort Tourgis, and their guns would have
been trained on invaders landing at either Saline or Crabby
Bays. Fort Platte Saline at its strongest only mounted five
68pdrs, while Fort Doyle was equipped with two 8in guns.

Today Fort Platte Saline can hardly be identified, for it forms
part of the works of the Alderney Metallic Grit & Gravel Co
Ltd, whose activities include the excavation and export of
beach material nearby. Fort Doyle, named after Lt-Gen Sir
John Doyle, perhaps the most distinguished Lieutenant-
Governor the Bailiwick of Guernsey has ever had and its
Commander-in-Chief from 1803 to 1816, has enjoyed practically no history (like its fellows), though the Germans saw fit
to fortify it anew during the occupation.

Fort Clonque

Perhaps the most picturesque of Alderney's defences is Fort
Clonque, standing beneath towering cliffs on a rocky islet
almost at the western mouth of the Swinge and covering its
approaches; and it is well nigh invisible from the sea. A causeway connects the fort with the shore. When it was built in 1854,
its barracks were on the landward side of the causeway, which

has the disadvantage of being covered for a short period around high water.

In 1886 Fort Clonque mounted one 68pdr, four 64pdrs and one 8in gun, which covered any possible landings on the inhospitable shores of Clonque Bay. The fort itself was the last in the chain of defences in western Alderney; beyond it are high steep cliffs. Today the fort has been turned into holiday flats and is the property of the Landmark Trust, which bought it for £19,000 in 1965.

Fort Grosnez

Probably this is the oldest of the Victorian forts, for its prime purpose was to defend the so-called 'harbour of refuge' at Braye and its position is practically adjacent to the breakwater. Although of no great size, it was heavily armed, mounting in 1886 no less than fourteen guns in eight batteries, several of them being of 8in calibre. Like many of its fellows, the fort was manned by the Royal Garrison Artillery, and in the lobby of the Court House hang photographs of these troops manning the fort. However, it was not used by the military for many years, and was chiefly employed in housing equipment for the maintenance of the breakwater, a purpose it still serves to this day. Within its walls of finely dressed stone are stores, workshops, garages and offices. Signs of German occupation are prominent, needless to say.

Château à l'Etoc

This name suggests the presence of a fort long before the days of Queen Victoria, yet there appears to be no record of such an ancient stronghold. The name means 'Castle of the Stack' and may well have come from a rock formation resembling a fortress. The site was of prehistoric importance, as we have already said (p 44).

The fort stands on a low headland separating the bays of Saye and Corblets. Not far off is Fort Albert. The barracks are handsome and, from the land, mask the gun position seaward of the building. But German fortifications have wrought havoc with the Victorian work, which included the Hermitage Battery, mounting a 68pdr, and six other batteries equipped with 8in and other guns. Today the fort has been turned into a block of flats, and its good looks match those of its fellow, Fort Corblets, across the bay.

Fort Corblets

In 1886 this fort mounted six guns, the heaviest being 64pdrs. Now a private residence, Fort Corblets was built of the decorative reddish local sandstone and, like other forts, its elegant mass merges well into the scenic background. It stands on the site of a former battery, built on Grande Corblette, which mounted two 9pdrs.

Fort Corblets seems to have enjoyed a short life as a military building and, in the early part of this century, was in poor shape. At one time it was used as a tea house; then it became the summer residence of the late F. L. Impey. After the German occupation (when the Germans made use of the fort), Mr Impey thoroughly restored the place, without robbing it of its martial appearance, and transformed it into a residence of the utmost comfort and charm.

The Eastern Forts

The now ruined Fort Les Hommeaux Florains poses dramatically on what was once styled the 'Islet of Flowers', within a stone's throw of Alderney lighthouse at Mannez. In its prime the islet, covered by Victorian fortifications, was linked with Alderney itself by an artificially constructed causeway, traces of which survive at both the landward and seaward extremities.

It is difficult to imagine an enemy seeking a landing at so inhospitable a situation, Fort Hommeaux Florains mounted no more than two 32pdrs and a 68pdr. It must have been garrisoned for but a short while. The story goes that it was once used as married quarters, but the wives, dismayed at its desolate setting, refused to remain there. The building was suffered to fall into decay, and even the Germans ignored it. Its sole claim to fame lies in the wreck of the sailing ship *Liverpool* on the eastern rocks of the islet.

Another neighbour of the lighthouse at Mannez is Fort Quèsnard, whose cannon must have covered the waters where the Swinge and the Race of Alderney meet. The building is relatively undistinguished and, even when in commission, merely mounted four 8in guns. The Germans fortified the place and, of course, did nothing to improve its appearance.

More attractive is another islet fortress, Fort Houmet Herbé, overlooking the Race and sited between Quèsnard and Longy. A causeway linked it with the foreshore and its remains survive. Again, with the savage Brinchetais reef so near, one would have thought a fort hereabouts superfluous. Its three 68pdrs once covered the rocky approaches to Grounard Bay, the only possible place for a landing, and their mountings are still to be seen. Like its contemporaries, this fort has much finely dressed granite, with an impressive entrance. The Germans seem to have ignored it.

The Longy Defences

Rather rambling is the attractive fort on Raz Island, at the entrance to Longy Bay. It is reached by causeway and its guns—originally four 64pdrs in two batteries—helped those of Fort Essex to cover the approaches to this broad sandy bay, so inviting to the invader. The fort itself does not seem to have been used over-much by the Army, though the Germans fortified it and took it much more seriously. The best thing they did to Raz Island was to improve the causeway, to the satisfaction of those who, following the liberation, turned the fort into a private residence.

Across Longy Bay is Frying Pan battery, so called because of its shape. This was part of the Victorian defences styled 'Longy

Lines'. They extended below the hillside whereon stands Fort Essex (now styled Essex Castle) and at Frying Pan battery itself there was once a 64pdr. Three 8in guns occupied a neighbouring battery and a 10in howitzer was mounted in another nearby.

Despite its name and the impressive position it occupies, Fort Essex does not appear to have been armed with guns by the Victorians. It served for years as a military hospital, although the Germans fortified it, which is not surprising, when one considers its site on a hill dominating so much of eastern Alderney and the Race.

THE GLORY FADES

In 1884, when a Memorandum was published dealing with the defences of the Channel Islands, it was stated that the objects for which the naval harbour at Alderney were designed had by then lost their significance. Steam had changed naval strategy and Alderney, like the other Channel Islands, was of far less importance than in the days of sail. To have remodelled Alderney's defences would have cost £75,000, even more if guns fitted to cope with modern warships were installed, and 'since there is now nothing to defend in Alderney it may probably be taken for granted that no such sum will ever be spent'. It was recommended that Fort Albert be the principal source of defence and that 'the remaining works be dismantled and rendered indefensible'. It was suggested that their sites, if sold, would realise £9,200, 'and the conditions of sale might include demolition'.

Although Fort Albert was, indeed, made the chief defence work, the other forts were not destroyed or sold until very many years had elapsed. The big arsenal at the foot of Fort Albert was retained in commission, though neighbouring Mount Hale battery was no longer used. Some of the married quarters in the region of Whitegates continued to be owned by the War Department until well after the German occupation.

7 FROM QUEEN VICTORIA TO HITLER

SO many workmen were engaged on the building of the forts and breakwater that many houses were built in the districts of Crabby, Braye and what came to be called 'Newtown', a settlement near the excellent school, built in 1969 between Braye and St Anne. The new church of St Anne was built, as was Scotts Hotel in Braye Road, Alderney's premier hostelry, kept by Mrs Scott, wife of Captain George Scott of the packet *Queen of the Isles*. There were other establishments, too, including the Belle Vue Hotel on Butes, and the inns at Braye.

A post office was established in Alderney in 1843 and at about that period a new Court House was built in the town. This stands in New Street, formerly Rue des Heritiers. In 1850, following the consecration of St Anne's Church, St Michael's cemetery, which was the former burial ground for strangers on the Longy road, was dedicated by the Bishop of Winchester. At the close of the nineteenth century a new and controversial jetty was built at Braye, and among the shipping using it was the local steamer *Courier*.

THE BOER WAR AND WORLD WAR I

In 1905 the Duke of Connaught visited the island, following a brief stay in Guernsey, and unveiled the South African War memorial. Seven Alderney men died in that conflict. The Duke inspected military establishments, was cordially entertained and duly left in the cruiser *Monmouth*. In honour of his visit St

Anne's Square was renamed Royal Connaught Square, as a plaque declares on the vicarage wall. Not far away, in a little garden in Victoria Street, the Alderney memorial records the names of forty-two of her sons who were killed in World War I and twenty-four in World War II.

During World War I the island was no longer regarded as the 'Key of the Channel', yet it still had a few modern guns at Fort Albert and some troops. The regulars remained in the Channel Islands, although by 1917 they had been reduced to the Royal Defence Corps. Throughout the war the Militia assisted them, though the younger islanders served overseas in the Royal Guernsey Light Infantry and other units.

Sometimes enemy activity off Les Casquets presented a hazard to the little inter-insular steamer *Courier*, though French seaplanes, based in Guernsey, kept a vigilant watch for U-boats. The *Courier* was painted a naval grey, but carried no armament.

BETWEEN THE WARS

The war over, Alderney resumed its peaceful life. Farming (including the breeding and export of cattle) and quarrying ranked as its most important industries. It was self-supporting in milk, butter, pork and eggs, and farmers grew sufficient winter fodder for their animals. There were about forty farms, the largest being 20–25 acres in extent. Much of the farmers' land might be scattered all over the island and, thanks to its complicated inheritance laws, the strip system of agriculture sometimes resulted in about 25 acres being divided into some thirty-five plots, which was anything but economic to administer. Yet this archaic arrangement persisted, because the Channel Islander is conservative.

Quarrying was mainly in the hands of the Channel Islands Granite Co Ltd, a subsidiary of Brookes Ltd, of Halifax. It owned some 400 acres of land and leased a further 280 acres from the Commissioners of Crown Lands. This body owned a considerable area of the island and over further regions the War Office had manoeuvring rights; in the main these areas were otherwise of no great value. The quarrying company leased the Admiralty railway, together with quarries and the foreshore at La Cachalière, on the south coast, where a quay was built early in the present century for the export of stone quarried close by.

The company employed about 100 men and boys and was the largest employer in the island. The annual wage bill was about £14,000, and the average man's wage was £2 per week. The majority of quarry workers lived at Newtown, Crabby and the Braye neighbourhood. Between the wars, quarrying was undertaken at Mannez, the Battery quarry (inland from Braye Bay) and York Hill (below Butes). Work at La Cachalière was abandoned because of its navigational hazards. Gravel was excavated at Platte Saline, as it still is.

In the 1930s, Alderney had its own journal, the *Alderney Weekly Times*, edited and published by a Scot, Ian Glasgow, of forthright views. There was also a town crier, who existed until comparatively recently; he wore no uniform but used the traditional bell. There were a few motor vehicles in the island, including some rather primitive buses, but most of the transport was horse-drawn. Gas was available in St Anne (the works were at Newtown and the present Harbour Lights Hotel covers the site) and, in the 1930s, electricity was introduced. There was a telegraph service, but no telephones, apart from a solitary radio-telephone.

Alderney enjoyed a small tourist industry and there were numerous hotels and guesthouses. Excursions were frequently run from Guernsey and occasionally from Weymouth and Poole. There was a cinema, 'The Rink', in High Street, dances were sometimes held at the Victoria Street Assembly Rooms (now occupied by business premises) and at the Militia Arsenal. Music was provided by the Alderney Orchestra, the band of the regiment on occasion, and by that of the Salvation Army.

Sport added to life's enjoyment, including football, cricket and tennis, while golf was played on a course below Fort Albert. Fishing was as popular then as now.

Alderney was the first of the Channel Islands to have an airport, opened in 1935 and controlled by Wilma Le Cocq, an island girl. There were no precise runways and aircraft used a flat grassy stretch of what used to be the western end of the Blaye. The terminal building was a corrugated iron shed and conditions within were primitive. The airport was primarily built for the transport of guests to and from the Grand Hotel, standing on Butes. It provided a summer service only, and functioned until the outbreak of World War II, which compelled its operators, Jersey Airways, to suspend the service.

THE CONSTITUTION

Alderney's prewar local government (indeed, up to 1949) comprised the Court, States and *Douzaine*. The president of the first two was the judge, a Crown appointment. The Court dealt with civil and criminal cases, while the States was responsible for the maintenance of public services, the framing of *ordinances* (local laws), finance and the collection of dues. The *Douzaine*'s task was limited to parochial duties, broadly speaking, and its president was styled the Dean (*Doyen*).

The Court comprised the judge, six jurats (magistrates), the *procureur du roi* (Crown officer), a *greffier* (clerk), a sheriff and a sergeant, the last two being responsible for carrying out the Court's orders. The jurats were elected by ratepayers and were unpaid. Like the judge, the procureur, greffier, sheriff and sergeant were appointed by the Crown. The States consisted of the Lieutenant-Governor of Guernsey (or his representative), the judge, jurats, four *douzeniers* (members of the *Douzaine*), the Court officials and three people's deputies, elected by public suffrage and holding office for 3 years.

There was also a Court of Chief Pleas, which met in January

and at Michaelmas. It comprised the judge, jurats, twelve *douzeniers*, Court officials and four *connétables* (constables), who were elected triennially. The Court's business was to deal with matters of general concern, and members of the public had the opportunity of addressing it on pertinent subjects. All these assemblies were held in the Court House, and here was to be found the Crown receiver, who represented the Crown Lands Department and was appointed by the Crown.

8   SWASTIKA OVER ALDERNEY

WHEN war broke out in September 1939, the garrison in Guernsey was withdrawn and, at first, there were no troops in Alderney, though the Royal Guernsey Militia was mobilised to defend its island. However, late in September 1939 the first of the military arrived in Alderney to form a machine-gun training centre. They travelled in the *Courier*. Officers were accommodated at the Grand Hotel and other ranks at Château à l'Etoc because, at the time, Fort Albert was in the throes of rehabilitation. The staff at the centre ultimately numbered 200, and in all 1,000 troops were eventually stationed at Fort Albert, Fort Tourgis, the Arsenal and Coastguards, the two last being establishments near Fort Albert.

Thousands of pounds were spent on modernising these buildings, for the Army had no inkling that the troops' stay in Alderney would be a brief one. Yet in mid-June 1940 the centre was shut down, in view of the military situation in France. The faithful *Courier* was employed in the removal of the troops, and the guns and stores were loaded on the SS *Biarritz*, which conveyed them to Southampton.

At this period the civilian population totalled about 1,400. There were also over 500 cattle and forty horses, as well as sheep, goats and pigs.

In the early days of June 1940 France's resistance to Germany was crumbling, the British were leaving Dunkirk and other parts of the continent and seventy men, women and children of Boulogne resolved to find refuge, if only temporary, in the Channel Islands. They travelled in three fishing boats and, during their brief passage, two died. They, like others, thought they would be safe from invaders in Alderney, the first island they arrived at, but their stay there was brief. True, the penniless refugees were made welcome, but the French people soon discovered that the islanders themselves were about to become refugees and so, with sad hearts and following a brief rest, they went their way.

In those anxious days confusion reigned, not only in Alderney but in the larger islands, if not in Britain itself. Defeat stared the Allies in the face, the invasion of England appeared imminent and the Channel Islands, far from being sanctuaries, seemed likely to be stepping stones to be used by the victorious Germans. In Guernsey and Jersey the evacuation began of men of military age, schoolchildren and, indeed, of all who wished to leave, but in Alderney no arrangements seem to have been made. The telegraph cable to England was out of order and radio silence, imposed by the Government, ruled out communication in this respect. What few messages passed between Guernsey and Alderney were taken by fishing craft, for the steamer service had become increasingly irregular.

As the Germans advanced, a mighty pall of smoke arose from Cherbourg and, because there was little wind, it drifted slowly out to sea and eventually swept over the south coast of England. It caused uneasy speculation in Alderney and those who suspected it was due to German destruction at Cherbourg were correct. More refugees tried to land in Alderney, but were sent on to Guernsey. The lifeboat from Cap de la Hague put into Braye and its crew and the French troops aboard reported that the shores of Le Cotentin were swarming with Germans.

At that period Judge F. G. French was virtually in command in Alderney, and his policy decisions at that testing time have been severely criticised. Nevertheless, the situation was obscure and patently dangerous, and whatever resolutions were passed, they were certain to displease some. Thus when Nicholas Allen, chief pilot and acting harbour-master, reported seeing hordes of Germans on the French coast, Judge French declined to adopt his suggestion that Trinity House should be advised of the position, despite the radio silence.

Mr Allen, who was Trinity House pilot in Alderney, decided to send a message to the acting postmaster who, on his own initiative, despatched it the following day. Trinity House replied, on 19 June, stating that the *Vestal* was on her way to remove the keepers at Les Casquets and Alderney lighthouses. On the same day the steamer *Sheringham* arrived from Guernsey to evacuate those desirous of leaving Alderney, but she had not been expected and it was thought she was there to remove men of military age and children, as was happening in Jersey and Guernsey. A few people left by this vessel, but the majority remained.

Judge French wrote to the Lieutenant-Governor of Guernsey and Pilot Jack Quinain was asked to deliver the letter in his boat; but, unknown to the Judge, His Excellency had been recalled to England, because of the military situation, and the letter reached the Crown Officers of Guernsey instead. They immediately telephoned the Home Office for vessels to be sent to evacuate those Alderney people anxious to leave. By then the *Vestal* had arrived at Braye.

Meanwhile, Judge French convened a States Meeting and the town crier was instructed to advise all to meet on the Butes to hear what had been decided. The Judge said he had asked the captain of the *Vestal* to inform the Admiralty that the islanders were prepared to remain unless it was thought Alderney would be invaded. He displayed the following notice to the public, which has been preserved in the Island Museum. It was dated 22 June 1940 and bore his signature:

I have appealed to Admiralty for a ship to evacuate us. If the ship does not come, it means we are considered safe. If the ship comes, time will be limited. You are advised to pack one suitcase for each person, so as to be ready. If you have invalids in your house, make arrangements in consultation with your doctor. All possible notice will be given.

He asked the people of Alderney, 'Do we go or do we stay?', when they assembled at Butes. They unanimously voted in favour of leaving and the Judge told them that when the ships were sighted the bells of St Anne's church would be rung and all must make for the harbour forthwith. A close watch was kept and on Sunday, 23 June, the ships were observed and the bells rung.

The tragic exodus is graphically described in *The Alderney Story*, published by the Alderney Society and written by Michael Packe and Maurice Dreyfus. The ships took the people to Weymouth and on one of them three babies were born in mid-Channel. At Weymouth the party was given temporary shelter and food, before leaving for different destinations.

Alderney was not left entirely deserted. A few seafaring men remained to perform final duties and there were, as well, nineteen islanders who refused to leave. Ultimately, twelve were persuaded to move to Guernsey but seven would not budge. The island's 500-odd cattle were taken to Guernsey, together with twenty-three horses and about thirty pigs. The local ships *Courier* and *New Fawn* were employed on this mission, and on her return to Guernsey the *Courier* ran into the German air raid on St Peter Port. However, she escaped attack and next morning sailed for Plymouth with passengers and Alderney's pigs still aboard.

THE OCCUPATION

On 2 July 1940 the Germans landed at Alderney airport. More came by sea from Cherbourg. The garrison of only eighty, was first commanded by a Sergeant-Major Schmidt, but he was replaced within a week or two by a Sergeant-Major Koch, and

Page 89 St Anne's Church is deemed by many the finest of the Channel Islands' modern churches. It was built in 1850 and replaces the more ancient building, whose clock tower survives at Le Huret. The island church stands off Victoria Street

Page 90 (*above*) Sauchet Lane, an old region of St Anne. Cobbled streets and fine old stone-built houses give this and other streets both charm and character. Little traffic disturbs its ancient peace; (*below*) Victoria Street, seen from halfway down its incline. During the summer it is a pedestrian precinct but at other times most of the thoroughfare is a one-way street, due to Alderney's increasing traffic

a party of Guernsey workers augmented the seven remaining civilians in putting the land to rights and generally tidying up. Pilfering was rife, however, and not only Germans were involved. The old steamer *Staffa* resumed the link with Guernsey, together with the small motor vessel *White Heather*, formerly on the Guernsey–Brecqhou service. There was also a craft named *Spinnel* in use. In 1941 the *Staffa* was wrecked at Braye, and her rusty remains are still visible near the jetty.

In 1941 the Germans resolved to make Alderney the granary of Guernsey, and it was planned to cultivate 700 *vergées* (2½ to the acre) on the Grand Blaye and 300 on the Little Blaye. The crops would comprise barley, oats and wheat. Some of the island herd was returned from Guernsey, and more workers from that island arrived and were quartered at the Grand Hotel with the others.

The Admiralty breakwater requires constant maintenance, but for a year the Germans did nothing about it, until the Guernsey authorities finally convinced them of the importance of attending to it. The Germans also extended the jetty by building an iron pier, whose substantial remains today pose a problem to the States, for the structure is now both unsafe and useless. Builders of this pier were civilians who had just finished La Maseline harbour, in Sark. They lived at Fort Albert.

By April 1941 the garrison was due to be strengthened and to this end houses in St Anne were repaired, by more workers from Guernsey. St Anne's church was desecrated by its use as a wine store and, later, as a butcher's shop. Looting and damage to property continued. Subsequently several Guernsey workers were convicted of these crimes and sentenced in the Guernsey Magistrate's Court.

On 20 October 1941 Hitler ordered the fortification of the Channel Islands on a grand scale, in order to prevent the British retaking them. The Germans regarded the islands as of importance in the protection of their sea communications, and it was decided to station a division of troops on them.

German camps were set up at Borkum (off the Longy road), Helgoland (inland from Platte Saline), Nordeney (near Saye Bay) and Sylt (south-west of the airport). In 1942 the Guernsey civilians were sent home, to be replaced by others, far less happy. They were the unfortunate workers for the Organisation Todt who, by the thousand, were employed under slave conditions to build the German fortifications. They were treated harshly and many died in the island. They were of varied nationalities, and those Russian workers who perished were buried in a special enclosure on Longy Common.

The German garrison numbered 450 in 1941, but by 1944 it had increased to 3,200, comprising infantry, navy, air force and other services. It is on record that 37,000 mines were laid around the coast and a particularly stout anti-invasion wall, which still exists, was built at Longy Bay. Guns were mounted in batteries. They included cannon of large calibre as well as AA guns. The airport, however, was not used for fighters, contrary to reports.

Mannez lighthouse was not occupied by Germans, but Les Casquets was, and in 1942 a British raid on it resulted in the capture of its occupants. The raid was led by Major March-Phillips, and Alderney-born 'Bonnie' Newton also took part. Five nights later commandos landed at Burhou, but found it deserted.

The German fortifications were known to the British authorities long before the war ended. The Army issued a map of the island showing the defences in the greatest detail, and obviously this information had been derived through the medium of intelligence bulletins and sorties by RAF aircraft fitted with cameras. The map revealed that the breakwater, piers and harbour sides were equipped with mines. Barbed wire festooned the slipway and other possible entrances, there were ubiquitous booby traps, underwater obstacles and tank traps were set in various strategic positions, while bunkers and pillboxes abounded.

On the hill above Mannez stands a massive German tower, once dubbed 'The Odeon'. Such bays as Braye and Longy had booms extending across them, and at the feet of the cliffs mines were hung. Great ammunition dumps were established. There were transport bays and rough camps for the workers, but the troops, naturally, were better housed.

A cemetery for the German dead stood near St Michel's graveyard, on the Longy road, but after the war the remains were taken for reburial in France. The Germans turned the Roman Catholic church at Crabby into a store. A further sign of their occupation was the substantial tower, still standing, which they built at Les Mouriaux, in St Anne. They also built their own power stations, and there were underground chambers, a telephone system, lazarets, coal stores, kennels and other necessities to an island which had been turned into a second Gibraltar.

There were stables for the fine horses the Germans imported. The horses were far better cared for than the wretches who built the defences. They were starved, badly clad, overworked and brutally treated. Some, including Vichy Frenchmen, fared better than others, but Russians and Jews suffered most. The maximum number of slaves is said to have exceeded 7,000, of whom thousands died.

HMS *Rodney* shelled the battery in the centre of the island in June 1942, her 16in guns firing at 20 miles range for over 2hr. The battleship's attack appears to have marked the start of the island's decline as a place of military importance.

Meanwhile, in the United Kingdom, many of the people of Alderney had moved to Glasgow, though some went to Birmingham, and others to London, where the Channel Islands Refugee Committee had its headquarters. An Alderney Relief Committee kept in touch with islanders, and many leading Alderney people served on it.

GERMAN CAPITULATION

On 16 May 1945, a week after the Germans had surrendered Guernsey, they capitulated in Alderney. A British force, part of Guernsey's liberating troops, went to the island and accepted their surrender. It could hardly be styled a 'liberation', for the population were still absent. The island's condition was so deplorable that the Home Office set up a Commission of Enquiry, which made plans for its rehabilitation.

A vast amount of work needed to be done and much of it was carried out by prisoner of war labour. Mines and other lethal weapons had to be removed, the place had to be cleaned, buildings repaired and land made fit for planting. A hundred and one jobs awaited attention. Alderney in 1945 had never looked more devastated, and there were some who declared it would never recover.

Stories of Nazi atrocities in Alderney found their way into the national press soon after the war in Europe ended. There were tales of gas chambers, mass extermination of slave workers, of the island being a prison fortress. Investigations into these allegations were promised, but there were difficulties. The workers had been removed soon after their task was completed, and the Germans who surrendered could scarcely be expected to give evidence, even if they had been aware of all the facts. So no official investigation followed into the alleged war crimes in Alderney. The unhappy past was pushed aside, to make way for the major enterprise looming ahead.

9   RENAISSANCE AND ITS AFTERMATH

ONE man who played a prominent part in Alderney's re-habilitation was the Lieutenant-Governor, Lt-Gen Sir Philip Neame, VC, who was appointed in the summer of 1945. He should be best remembered for his unflagging efforts to put Alderney on her feet again. It was natural that, when the Home Office appointed a Commission of Enquiry to study Alderney's future, he should be its chairman. There was, at one time, a notion that Alderney, at least temporarily, should be abandoned. Happily, this policy of despair was not adopted.

It was decided that the islanders should return in phases. An advance party would include States members and business people, after whom would follow those prepared to work on the land. Shopkeepers would take no money but receive a standard wage, and would be provided with essential equipment and an initial stock of wares. Before this could happen, dwellings and shops must be put in order and about 300 of these were to be dealt with by a special labour force. The Ministry of Works erected several houses that today stand as agreeable reminders of ministerial good taste.

Were the people of Alderney prepared to go back under such conditions? All were circulated during that autumn of 1945 and the majority answered 'Yes'. The Home Office set up an Alderney Resettlement Section at Whitehall, and its task was the despatch of necessities to the island and arranging the return of its inhabitants. The Alderney Relief Committee suggested who should return to the island first.

By October 1945 the composition of the advance party was

finalised. The WVS played a major part in catering for the welfare of the returning islanders, working closely with the Home Office in doing everything possible for their comfort. By this time 300 houses were habitable, utility furniture had been supplied and the former convent (now the Island Hall) had been turned into WVS headquarters. It was planned that the first arrivals should stay at the Grand, Belle Vue (now Bligh's) and Victoria Hotels.

The exiles returned by sea (the airport was still out of commission), travelling to Guernsey and thence to their island. The first party, which included Judge French, arrived home on 2 December, and a second party 2 days later. Both travelled on the small Government vessel *Guillemont*, a craft much used by Lt-Gen Neame. The weather was bad, but this did not deter her skipper, Captain Hubert Petit, nor the courageous civilians aboard her.

As the little ship drew alongside the Braye jetty, a pleasing sight met the eyes of the first party. A huge placard was displayed bearing the words 'Welcome Home'. It had been made by German prisoners-of-war on the orders of the Lieutenant-Governor. A contingent of troops was on the pier to greet the party, who were taken to lunch at the Grand Hotel. Then a service was held at St Anne's church, conducted by an Army chaplain, though the building was in a sorry condition. No bells rang, for but two remained, and these hung from frames in the churchyard. The remainder had been sent to France by the Germans, but fortunately they were found at Cherbourg and the Army restored them to Alderney.

The days passed and witnessed more arrivals. Supplies, too, were coming in, rehabilitation work proceeded with a will and, after what must have proved an austere Christmas, Alderney faced the New Year with a glimmer of hope. The States, with the Army authorities, administered the island and Alderney people were given such positions as fuel controller, commodities controller, storekeeper and farm manager.

It was decided to have a communal farm rather than a number of farms, and for a time this was moderately successful. But islanders are not really communally minded in such matters, and later the old order was restored. At first two herds were kept—Jerseys (a legacy of the Germans) and Guernseys. It is illegal to keep other than Guernseys in the Bailiwick, however, so the other animals were at first segregated from the rest and finally exported to England. Pigs were kept and, temporarily, sheep, another occupation 'left-over'. The German horses were worked, and much ploughing and planting were undertaken, particularly on and around the Blaye.

By February 1946 the airport was open again and Rapide aircraft were operating a service. Gradually other amenities were restored, not least of which was the school. The Alderney Grit & Gravel Company resumed operations in the spring and thus gave some employment. Market gardening was introduced by F. L. Impey at the Valley Gardens, near the Terrace, in St Anne, but shipping had not yet resumed its prewar service, and this held up exports.

The Southern Railway's vessel *Autocarrier* had been sailing weekly from Southampton to Alderney between December and May, but this service was discontinued and landing craft from Guernsey were then run, though each could carry but twelve passengers. A weekly cargo service by the Great Western Railway from Weymouth helped matters, but again only a handful of passengers could be carried.

In the restoration of Alderney excellent cooperation was forthcoming from the prisoners-of-war, who worked well and were courteous and friendly. Civilians and troops, for their part, treated the Germans with consideration. Just before they left the island, on 1 June 1946, the Germans gave a farewell concert at the Lyceum cinema. At the end of that month the British troops departed and Alderney, bereft of these supports, confronted the future with rather mixed feelings.

97

TROUBLES OF THE REOCCUPATION

It was inevitable that, after the initial enthusiasm of the islanders' return, an element of dissatisfaction should arise. For a time they had been liberally helped by the British Government, but this could not continue indefinitely. There was also unrest at the way in which Alderney was governed. Judge French was an unpopular figure. Some of his ideas were unacceptable and an example of this can be found in the matter of furniture allocation.

A good many objects remained to be claimed by islanders, and their family treasures (which had been removed from the houses during and following the Occupation) were accordingly deposited on Butes. At the blowing of a whistle a concerted rush was made on the goods and those who could lay hands on any items could claim them, irrespective of whether they had originally owned them or not. Known as 'The Battle of Butes', this operation resulted in great dissatisfaction, and to this day such an odd method of disposal is a sore point among many.

But unrest went further than this. Many who had lived in Britain during the war had enjoyed benefits unknown in Alderney. In addition, some newcomers to the island were preaching the gospel of change. Reform was in the air!

It did not come solely because of the wishes of the Alderney population. The United Kingdom had advanced the island £165,000 'to lay the foundations of a prosperous future', to quote the words of the Home Secretary of the time, Chuter Ede, but this future could only be ensured if 'the administration of the island can be placed on a sounder and broader basis'. The States was asked to submit proposals for reforming Alderney's constitution, finances and public services, and providing better cooperation in farming and the marketing of produce.

The next event was the appointment of a Committee of the Privy Council in July 1947, with the Home Secretary as chairman. This body, which visited the island to take evidence, was watched with enormous interest, not only in Alderney but in the other Channel Islands, especially Guernsey. It spent 3 days in the island in September, hearing evidence from the public in the Militia Arsenal, in Ollivier Street. In England it heard evidence from officials of the Home Office, Ministry of Civil Aviation, Post Office and the Office of the Commissioners of Crown Lands.

The States submitted proposals regarding constitutional and economic matters, as it had been asked, but the Home Secretary was dissatisfied with them and revisited the island in January 1948. There he again met the States, together with Guernsey representatives. Alderney itself was gravely disturbed by this enquiry, for by no means all were in favour of radical change, although it was clear that some change there must be. Reform was not merely the whim of 'revolutionaries', it was the desire of the British Government, which Alderney could not ignore, even had it so wished.

States meetings and public assemblies were held and all manner of complaints and suggestions were voiced. In Guernsey the problem of Alderney's future was gravely considered and, as the British Government had resolved to bear Guernsey's war debt of £2,000,000 or more, its Bailiff, Sir Ambrose Sherwill, deemed it just that Guernsey should assume responsibility for Alderney's anticipated overdraft. In return, he thought, Guernsey should have some control over Alderney's administration. Sir Ambrose did not at once reveal these sentiments to the States, but later they were substantially adopted.

The Privy Council Committee's recommendations were published in a White Paper in October 1949, and presented by the Home Office. Guernsey was to assume financial and administrative responsibility for Alderney's airport, education, health services, immigration, police, major roads and sewerage, social services and water supply. Guernsey taxes, duties and water rate should apply in Alderney, whose laws would be modified

99

accordingly. No major expenditure in Alderney was to be incurred without the sanction of the Guernsey States. In that body two seats would be occupied by Alderney representatives, who would also sit on the Inter-Island Advisory Council.

The States of Alderney would comprise an elected president and nine members. The office of judge would be abolished and the Court would consist of a chairman and jurats (the precise number was not specified) appointed by the Home Office. All Crown appointments would be abolished and the Court of Chief Pleas would likewise disappear. The Crown lands and dues in Alderney, excluding the breakwater, would be transferred to the States, in return for which that body would maintain the harbour and pay the salaries of local officials.

These recommendations were adopted by the States of both islands. Since the transfer of functions on 1 January 1949, and the resultant changes in administration, Alderney now controls its electricity, water, roads and sewerage, as well as paying for education, police and other necessities provided by Guernsey.

Alderney's last judge was Sir Frank Wiltshire, who held office for a brief period following the resignation of Judge French. The first President of the States was Commander (later Captain)S. P. Herivel, who filled this position from 1949 until his death in 1970. He it was who welcomed Princess Elizabeth and the Duke of Edinburgh when they visited the island on a summer's day in 1949 in the battleship *Anson*. The future Queen, among other engagements, planted a tree in Royal Connaught Square, near that planted by the Duke of Connaught in 1905.

THE NEW ERA

One of the results of the occupation was the disappearance of land boundaries. This was attributed to the Germans, who may well have found boundary stones and other limitations irksome to their military needs. Whoever may have been to blame, the marks had gone on the islanders' return and it was imperative they be restored as quickly as possible. To do so proved a formidable task, which lasted for years.

The whole island had to be surveyed and a map drawn accordingly, on which properties were plotted and numbered. A register was also prepared, but before this could be done ownership had to be proved and a great deal of time was spent by Captain S. Doll, the Government-appointed Land Commissioner, and his successors, in interviewing islanders, hearing evidence and visiting sites in dispute. Ultimately, the land was allocated as nearly as possible as it was in 1940, and today modern boundary stones mark the properties' limits.

Part of the western end of the Blaye was acquired by the States, when this body became the owner of the airport, and the property was considerably extended. Improvements included a small terminal building, situated rather precariously near the grass runway and wanting in the refinements the present building has, situated as it is at a safe distance from moving aircraft. The runway often became waterlogged, and there were other hazards, but despite all this, British European Airways and, later, Jersey Air Lines, provided a good service, using Rapides.

Other postwar developments included the acquisition by the States of the dairy at Le Val and the island water supply. The Alderney Light & Power Company was responsible for the electricity supply, and, as with water, used German installations. The telephone exchange (administered by the Guernsey States Telephone Council) was opened in May 1949.

The Mignot Memorial Hospital was soon functioning again. In those immediate postwar days it stood near the foot of Victoria Street, almost opposite the present Victoria Hotel. Another rather similar establishment to resume its functions was the Jubilee Home for old people, established in High Street in 1887.

In 1948 and for several subsequent years HMS *Alderney*, the submarine adopted by the island, was a most welcome visitor. Another source of pleasure was the revival of 'Alderney Week', a summer festival of sport and entertainment. The annual shows of the Royal Alderney Agricultural Society were revived; they were normally held at the Old Mill Fields, part of Burland's Farm, on the Blaye. Unfortunately, these shows now belong to the past.

Tourism made a slow return, for there were, initially, transport problems. Not many establishments were ready to receive guests, anyway, and the island was scarcely attractive to most visitors.

Once Alderney was on its feet again, however, it was 'discovered' by settlers, from the UK particularly, dissatisfied with postwar conditions and seeking somewhere quieter than Jersey or Guernsey. They bought existing buildings or built new ones, which resulted in a new element in the population and a change in the landscape. It also agreeably supplemented the island's coffers and led to better amenities, including transport. Other improvements included a telephone link with Guernsey and elsewhere, and the establishment of the Alderney library, an excellent collection located today in the Island Hall, but originally sited in the old Militia Arsenal. Later it moved to premises in High Street before transferring to the former convent building, after the States had acquired it for £8,000 in 1959, when it was transformed into the island's community centre.

The Germans had wrecked the Catholic church at Crabby, so the chapel of the Convent of Our Lady in Royal Connaught Square was used for worship until the present church of St Anne and St Mary Magdalene was built at the top of Braye Road. Its foundation stone was laid by Cardinal B. Griffin on 23 September 1953, and it was opened for worship on 7 September 1958 by the Bishop of Portsmouth. Its basement is used as a social centre. The ruined church at Crabby was subsequently razed and a dwelling occupies its site. When the convent was bought by the States, the nuns moved to Guernsey, their departure coinciding with the closing of their excellent school in Alderney.

Following the formation of the Alderney Society in 1966, it was resolved to establish a museum, which was opened soon afterwards in the basement of the Island Hall. This period also saw the founding of the Alderney Preservation Trust, a body whose aim was to acquire and preserve places of outstanding beauty and interest. After some years' existence it was merged with the Alderney Society.

In the first decade following liberation, various industries came and went in the island. For example, there was the manufacture of silencers at the Penguin Works at Newtown, the meat factory at the Arsenal, Fort Albert, and reference has already been made to the market gardening project in the Valley. All of them petered out after a few years, chiefly because of the difficulty of despatching their goods to the United Kingdom. Fortunately, considerable areas of land have been and still are devoted to the cultivation of spring flowers and vegetables, involving the use of Italian labour. Some of these men and their families live at Fort Tourgis.

Occupation Cemeteries

A small and beautifully kept German cemetery adjoined the St Michel burial ground on the Longy road. It was dominated by an imposing memorial stone, bearing a German inscription. This cemetery was in marked contrast with the Russian burial plot on Longy Common. Other Germans were buried in St Anne's churchyard, and the authorities were content to allow all these burial places to remain. However, in 1961, the German War Graves Commission exhumed the bodies of seventy servicemen and reburied them in Normandy, the remains of 384 Todt workers were taken to the same place, and the bodies of eleven French workers were reburied in their homeland. In that year also the States of Alderney set up a small memorial on the road to Saye Bay, near the site of a camp where French workers lived during the war, in the shape of a blue granite boulder, inscribed: *A la memoire des Français morts pendant la guerre 1939–1945.* It bore replicas of the Alderney flag and the Cross of Lorraine.

In 1966 H. C. Hammond, then a States member and well known businessman, drew attention to the poor state of the plaque, and offered, with his brother Jack and their sister Mrs 'Babs' Tinson, to enlarge the memorial so as to commemorate the dead of all the nations associated with the occupation of Alderney by the enemy. This would be in memory of their two brothers, victims of World War I. This offer was gratefully accepted, and the new memorial was officially opened by the President of the States, Captain S. P. Herivel, in the presence of British, French, Belgian and Russian representatives. In 1970 there was an even bigger ceremony, since nearly 100 former slave-workers had made a pilgrimage to Alderney aboard British and French warships to attend.

Royal Visits

If ordinary tourists were somewhat scarce in the decade or two after the war, royal visitors were relatively plentiful. The Queen and the Duke of Edinburgh paid a return visit, by air, in July 1957. Her Majesty presided at a special meeting of the States, laid the foundation stone of the present Mignot Memorial Hospital, near Crabby, and, among other functions, attended a garden party at Mouriaux House.

Queen Elizabeth, the Queen Mother, visited Alderney in 1963 in the royal yacht, *Britannia*, using the breakwater slipway as Queen Victoria had done. Other royal guests were the Duke of Gloucester in 1954, Princess Margaret in 1959, Princess Alexandra in 1968 and Princess Anne in 1972.

10 PEOPLE, CUSTOMS AND LEGENDS

THE people of Alderney may be divided into working and leisured classes. Some might declare that the former are the true Alderney people, the latter being little more than parasites, but it is not as simple as that. The inhabitants bearing French names and those whose forebears settled there during the period when Victorian building work overshadowed every other local enterprise are, for the most part, those who staff the shops and offices, man the farms and workshops, work on breakwater or harbour, fish the sea and perform other essential tasks.

Of the hundreds of postwar settlers, some play an active part in island government and other insular duties, if few are engaged in industry. Their wealth aids Alderney's economy: for instance, the money they spend in entertaining aids the States revenue, while their other expenditure provides work for the islanders. Without the 'rentier' Alderney would be much the poorer. He may well have settled in Alderney or some other of the Channel Islands because of heavy tax burdens elsewhere, but often the reason is that, in his view, the Alderney way of life is preferable to that of the United Kingdom. While he may not spend all his time in the island, it is his home, and he would be the last to try to alter its pattern.

SONS AND DAUGHTER OF ALDERNEY

Much has been written of *Judge French*, under whose leadership the people evacuated in 1940. He has been bitterly criticised for the decision to leave, yet people forget that his counsel was accepted quite voluntarily. And when the folk of Alderney returned from exile, the Judge failed to please them. They resented his military bearing and orders. Yet French was working with the military; he held the rank of Brigadier and he was a disciplinarian. He made wrong decisions, of course—the 'Battle of Butes' was one of them—but he did much towards resettlement, beyond doubt. He was generous, too, with his wealth and the Mignot Memorial Hospital was one of the institutions benefiting from his kindness.

French's most bitter opponent was *William Herivel*, of the Marais Hall Hotel and a member of the States. He was a true son of Alderney who fiercely resented French's admittedly autocratic rule. Indeed, it was largely Herivel's unending criticism that led to French's resignation in 1946.

Another leading figure in Alderney's resettlement was '*Nic*' *Allen*, island-born fisherman, Trinity House pilot and hotelier. Allen's minesweeper had led the British ships into Braye at the liberation, and thereafter he was a particularly prominent personality in Alderney's maritime life.

A Guernseyman, *Col F. W. Marriette* became intimately connected with Alderney before and after World War II, in turn as businessman, schoolmaster and postmaster. A jurat, he became Chairman of the Court, following service in the States. His work in the island's rehabilitation was outstanding.

Peter Radice, as Clerk of the States (and also for several years Clerk of the Court), was a familiar island figure from 1949 until his retirement in 1973. He did an enormous amount of work behind the scenes and was a mine of information. He came to the island on his retirement, only to find work which was even heavier than that he had left.

Mrs Catherine Bickerton was a 'settler' who strove to make the island a better place, and a worthy jurat. She was a pillar of strength in the local Women's Institute, as she was in the Alderney branch of the WRVS.

Page 107 (above) Royal Connaught Square, showing the Island Hall which was originally Government House but later became a convent; (below) Marais Square with the Marais Hotel. Now the bus terminus, but within living memory cattle drank from its trough and animals destined for overseas assembled in the square, awaiting the drive through St Anne on the way to the steamer

Page 108 (*above*) The old pier at Braye, with the bathing beach beside it. On the heights is the rim of Battery Quarry, with a television beacon on the skyline. Stone was once shipped from the lower part of the jetty, known as Douglas Quay; (*below*) Braye Bay seen from Butes, with Fort Albert beyond. In the foreground are the houses of Braye and Newtown. The highway skirting the shore is the Lower Road leading to eastern Alderney

Col Marriette's successor as Chairman of the Court, jurat *Charles Richards*, was another impressive Alderney personality. His interests were wide, embracing work and play, the British Legion and music, while his achievements on the Alderney Relief Committee in London during the war were invaluable.

In *Captain S. P. Herivel*, Alderney had a most worthy son. He did much towards the island's recovery, and his influence with Whitehall and Guernsey was a prime factor in its resettlement. He was Alderney's first President of the States and, fittingly, died 'in harness' at an advanced age.

One who also helped in the difficult postwar years was *Tommy Rose*, the famous aviator. As an hotelier and a States member he proved, in practical fashion, his love for the island, and his jovial personality and zest for public work earned him both affection and respect.

Finally *Sir Henry Gauvain* was of considerable assistance to the exiled population in Britain during World War II. Despite his work as a surgeon, he found time to become the first chairman of the Alderney Relief Committee, and he always showed the deepest affection for his island.

ANCIENT CUSTOMS

In Alderney the French element is weaker than in the other Channel Islands, but it still lingers on those rare occasions when *Le Clameur de Haro* is raised, though to style this 'French' is really a misnomer, for it is actually of Norman origin. It is a cry for help, and the word 'Haro' is said to be a corruption of 'Ha Ro', calling attention to Rollo, first Duke of Normandy. Others believe 'haro' is merely an exclamation, like 'hi!'.

The *Clameur* is invoked if real property is endangered. An islander may own a tree that overshadows his neighbour's garden, and the neighbour may start to chop it down. The owner could sue his neighbour, but meanwhile the tree would have been felled, so he might well raise the *Clameur*, thus causing an immediate stay in the proceedings.

The procedure is for the injured party to fall on his knees and to cry 'Haro, Haro, Haro! A l'aide mon Prince. On me fait tort,' thus begging the Prince's aid because he is being wronged. The 'Prince' could be today's Sovereign. The man then recites the Lord's Prayer in French, in the presence of two witnesses, and whatever work is in progress must stop. The *Clameur* has the force of law, and he who ignores it does so at his peril. The case is then reported to the Court and an action may follow. No work on the site must be carried out until the matter is settled, and if no Court action results, a year and a day must elapse before operations can be resumed. There are times when persons resort to this ancient custom in respect of matters not concerned with real property, but its use is then invalid.

Another surviving custom is 'Milk-a-punch Sunday', which falls on the first Sunday in May. Originally it was customary for folk to make their way to the places where cattle were tethered (for they were not permitted to graze loose and, even today, it is usual to secure them) and to milk the cows without their owners' leave. Rum and sugar were added to the milk, sometimes an egg was used, and the punch was enjoyed in the open air. Today, the custom prevails on licensed premises. Punch is served free, though only one glass per customer is usual, and as the period during which the punch is served is limited, one must be very agile in order to patronise more than one establishment.

LEGENDS

Some insight into life in Alderney in former days can be gained by sparing a glance at its folklore. Humanity, as well as stones, plants and animals, plays a naturally dominant role in the beliefs of yesterday, and while some of the legends woven into the fabric of island folklore should not be taken seriously, they have a value in being very often the genesis of history. The subject has been well studied in F. M. Picot's 'The Folk-lore and Customs of Alderney', published in the *Transactions* of La Société Guernesiaise for 1929.

The Hanging Rock is one of the outstanding features of the Longy scene and old Alderney folk used to call it 'Madame Robilliard's Nose', because she was very inquisitive. It was said that a witch, to keep her quiet, made her nose into a landmark. The rock figured in another tale, in which the devil, learning that Guernsey wished to acquire Alderney, advised the Guernseymen to attach a rope to the Hanging Rock and, by means of a rowing boat, tow it to Guernsey. Needless to say, even the devil failed to bring about this acquisition!

Another tall story concerns the so-called 'Nunnery', which was never a convent. A fort, it probably received its name from some frustrated soldier, as like as not, comparing its solitude with that of a nunnery. The story went that the ghost of a nun walked there, and that at midnight on New Year's Day 'a little man is said to come out of a cupboard in the tower of Essex Castle and walk three times round before retiring'.

A sandstorm which destroyed the ancient town of Longy was said to have been God's vengeance on wreckers. On the other side of the island, at Clonque, a white bull was said to haunt the road and Le Petit Val, and he who beheld it must be prepared for coming danger. Its home, apparently, was within a great rock, La Grosse, which was left by the sea but twice a year.

Legends are associated both with Les Casquets and Ortac. The former, it seems, took its name from a Jersey *seigneur* whose wife eloped with a 'rich *Aurignais*'. The husband, on discovering this, hurled his helmet into the air and it turned into the rocks which, stated the tale, 'became the fear of all passing ships'. Ortac, it was believed, was the home of a spirit who ruled wind and water; and, until early in the present century, it was customary for fishermen to stop in the Swinge and drink to the brooding spirit on the great rock.

Not far off, on Clonque beach, is a rock in the shape of a chair, and it was believed that a monk once wrestled with the devil here. The holy man vanquished him, then sank exhausted on to a rock, which at once turned into a chair for his greater comfort. Another rock, 'The Lovers' Chair', stood at Val l'Emauve, near the cliffs of Telegraph Bay. It was once the trysting place of a Guernseyman and a beautiful Alderney maid, who, when attempts were made to prevent their union, joined hands and leaped into the sea. This rock was destroyed during the German occupation.

Trois Vaux, the trinity of valleys between Hannaine Bay and Telegraph Bay, was once the place of burial for those who committed suicide and at night 'the moans of their troubled spirits are heard'. It was said that no trees would grow there until they found peace and, to this day, it is treeless. Other ghost stories are told of Raz Island and of the burial grounds on the Longy road.

Like other communities, the people of Alderney believed in witchcraft, and tales were told of witches dancing at full moon round the 'witches' stones at Essex and Clonque'. The people also believed in fairies, as the road in St Anne's known as the *Allée ès Fées* (Fairies' Alley) illustrates. Here, it seems, the sprites were wont to gather, and some of their other haunts were the small dolmen known in the past as *Rocque a l'épine* (Rock of the clearing), still standing near Fort Tourgis, and the dolmen, now destroyed, called *Maison ès Drôles*, at Longy.

Finally, Alderney grasshoppers used to be called *Les Chevaux de St Georges* (St George's Horses). Why this alleged patron saint of Guernsey should be associated with Alderney grasshoppers remains a mystery.

TOWN AND HARBOUR

'ALDERNEY: about 1,000 inhabitants who live altogether in a valley of about nine score houses, wherein they differ from the rest of the islands that have their houses scattered, perhaps necessitated to it to defend themselves from the invasions of former piracies, that in time of war frequented these seas.' So wrote Philip Dumaresq in his *Survey of Jersey* in 1685. While the population figure seems rather high, his theory that collective residence was a precaution against piratic landings may have been sound, although these could well have occurred in peacetime, too. A more likely reason was the proximity of the Blaye, for surely those who cultivated it would prefer to reside nearby, especially as water was available and there was a reasonable amount of shelter from the winds.

Certainly Alderney differed from its neighbours in this respect, and also because its capital was not a port, as are the chief towns of Jersey and Guernsey. Alderney and Sark both have their major settlements a mile away from their harbours, both of which are about 250ft below them.

ST ANNE

'La Ville, ou St Anne' was how the Chevalier de Beaurain marked the town on his map of 1757. Today residents always style it 'town' rather than 'St Anne'. It occupies a fairly central position in the higher part of Alderney, within easy walking distance of the Blaye yet not far from the beaches of Braye, Crabby, Saline and Clonque, whence *vraic* (seaweed) was brought for use as fertiliser. From times long past there must have been roads leading to these shores, as there certainly was between St Anne and Longy, Alderney's ancient port. De Beaurain marked a road leading to Clonque and Saline Bays and another, Rue des Sablons (Sandy Lane), running down to Braye and Crabby. The Longy road was prominently marked.

Dumaresq, writing in 1685, stated that there were two harbours—at Crabby and *Longis* (Longy)—but the first must have been simply a natural one. He wrote that 'there is also in a bay a peer formerly begun', referring to the medieval jetty at Longy. As Alderney was practically self-supporting until a century or so ago, its harbour could have been seldom used by vessels of any considerable tonnage.

St Anne bears a distinctly French appearance, though some say it resembles a Cornish town. Old it certainly is, yet apart from the clock tower (the remnant of the original parish church) no truly ancient building survives. Clearly there has been wholesale rebuilding and, stone being universally used, the result is as harmonious as if the little town had sprung up overnight.

Since Marais Square is the first part of the town proper the visitor sees as he drives from the airport, and because this region is among the oldest in the town, it is as well to make it the starting place of a tour of inspection. The trough still stands where it did when cattle used it, but it is empty and now only serves as a leaning place for those awaiting the bus, which starts from there. Adjoining the trough is the Marais Hotel, formerly the Marais Hall. There have been licensed premises here since about 1900. At one time laundry was washed in the spring at the corner of Little Street (formerly Rue des Vaches, or Cow Lane), which fed the fountain in the square.

Running westward is La Trigale, a narrow road in which stands a house with the only so-called 'Norman' arch in Alderney, although in the other islands this feature is common. The name Trigale means 'three ways'—Rue de la Fontaine (now Trigale road), Venelle Jehannet and Hauteville. The first two run from the west into the square, and the last is a street merging with St Martin's. Northwards are Les Mouriaux and Royal Connaught Square.

From Marais Square Le Huret leads towards the town centre. On the right is the Rose and Crown Hotel, and nearly opposite is the site of a smithy. Roughly east are the *venelles*, lanes connecting the town with the Blaye. Venelle des Gaudions has an offshoot, the diminutive Venelle du Milieu, and this doubles back to Le Huret. Beside the Rose and Crown is the Venelle Simon, very picturesque and steep. Off the High Street (Grande Rue) are the Venelle du Puits and Venelle du Sergent, near the Salvation Army building, once a Primitive Methodist chapel. Nearby is Sauchet Venelle, where a public pump once stood opposite the fine old Sauchet House.

Just beyond the Rose and Crown a hill runs down to Royal Connaught Square. This is Le Huret, from *hure*, a slope. Hereabouts was the ancient meeting place of the Court of Alderney, held in the open air. Even today royal proclamations are read there, among other traditional places. A plaque on a wall records the royal visit of 1957. The original Rose and Crown was located in a substantial house on the left of the slope leading to the square.

The reddish-brown setts of Alderney sandstone blend well with the buildings, none of which is of special architectural merit, until one reaches the former St Anne's Square, now called Royal Connaught Square. Here stands the Island Hall, formerly the convent and originally Government House. Its neighbour, the vicarage, is imposing in its way, like the Royal Connaught Hotel. Although the Hall is the most important building, it does not overpower the others.

Facing the Island Hall's forecourt is a fair-sized dwelling, flanking a lane running up to St Martin's. Here, from 1947 to 1964, lived the late T. H. (Tim) White, author and eccentric.

Running off the Square is the lane of Mare Jean Bott, full of interesting houses. In St Martin's is Les Chevaliers, an example of the farmhouses which once almost filled St Anne. Beyond the Square, north-west of the Island Hall, is Les Mouriaux House, one of Alderney's most stately buildings and the former home of the first President of the States, who bought it from the Le Mesurier family. Unhappily, it faces a hideous tall concrete tower, of German origin. Nearby is the Alderney Pottery.

Returning to Le Huret, one is confronted with the clock tower, all that remains of the ancient church of St Anne, standing in a fragment of its churchyard, where the Courts of the king and the bishop are said to have met. The precise age of the church is unknown, but the clock tower was built in 1763, thanks to the Le Mesuriers. In 1790 the church was improved by a gallery, built across its west end. The tower contains two bells, one inscribed 'Clement Tosear cast mee in 1701' and the other 'J. Le Mesurier praefect T. Le Cocq aedit anno domini 1760'. The bells strike the hours and the quarters.

According to A. H. Ewen's 'The Town of St Anne, Alderney', published by La Société Guernesiaise in its *Transactions* for 1958, the earliest reference to a parish church there was in a charter of 1054. Another of 1134 records the dedication of the church to St Mary. This name persists in subsequent records until some time between 1570 and 1660, when it became known as the 'Temple and Chapel of St Anne'. There appears to have been no precise reason for this change of name, though it is possible that the chapel was part of the original church. Mr Ewen stated that when the Le Mesuriers added a cross-aisle on the north side in 1761, 'this part of the church was known as the chapel'. He recorded that early in the eighteenth century the township itself became known as St Anne, before which islanders always referred to it as 'La Ville', as they do today, only in English.

In 1568 Alderney was transferred from the see of Coutances to that of Winchester. Its last priest was Pierre Charles and its first Protestant minister was Pierre Herivel, appointed in 1569 but soon succeeded by Edmond de la Rocque, instituted 'rector of St Mary's'. The term 'rector' appears occasionally in the Alderney parish registers, though 'ministre' does also. The holder of the living is now known as the vicar, the office having possibly been introduced with the building of the new church.

Until the mid-nineteenth century, most of the incumbents spoke French, as the islanders did, even though Alderney-French was dissimilar from that spoken in Paris. One of them, Jean Chretien Ubele, appointed in 1812, had his licence revoked by the Bishop of Winchester in 1818 on the grounds that he stole an altar cloth. Two years later the 'parsonage' was rebuilt, again through the munificence of the Le Mesuriers.

When John Jacob visited Alderney to gather material for his *Annals* of the Bailiwick of Guernsey, published in 1830, he recorded that a market house built by order of the States in 1799 stood in the square, near the churchyard, though it was only opened once during his stay there; and that a boys' school built in 1790 and endowed by Governor Le Mesurier stood nearby. Not far off was the Methodist chapel, built in 1813, which is now the Masonic Temple in Church Street, a thoroughfare linking the square with New Street. In Jacob's day this was 'the only place of worship for the dissenters in the island'.

High Street runs east from Le Huret and one of its most important edifices is the former school, endowed by Jean Le Mesurier in 1790 and now the museum of the Alderney Society. It was opened by the Lieutenant-Governor of Guernsey, Vice-Admiral Sir Charles Mills, in 1972 and houses an excellent collection of island treasures, among which are the Iron Age discoveries of its director, K. Wilson, made near the Nunnery and most carefully restored by his wife. The premises became available when a new island school was opened near Newtown in 1969.

Nearly opposite is the Coronation Inn, noteworthy for its sign depicting the crowning of Queen Elizabeth II. A little farther along High Street, but on the north side, is the Campania, an attractive public house. Not far away is the automatic telephone exchange, and almost opposite the Campania is Le Bourgage, an ancient part of the town, although the present buildings belong to a much more recent period. From Le Bourgage the road called Le Brecque leads straight to the Blaye. At its southern end another lane, Le Colimbot, takes one into Little Street, a further approach to the Blaye.

The High Street merges into Longy Road once the top of Le Val is passed. This valley runs down towards Braye, and parallel with it is the attractive Water Lane, styled Fontaine David or Val Reuters in its lower reaches. This emerges into Newtown.

The user of Longy Road passes Verdun Farm before reaching the St Michel burial ground, the old Strangers' Cemetery of 1806. St Michel's graveyard was consecrated by the Bishop of Winchester in 1850. It adjoins a Roman Catholic cemetery which, nowadays, seems seldom used. Above is the ridge of Les Rochers, with its conspicuous television mast. A little farther east is the nine-hole golf course, opened in 1970. Near the ninth green, and on the left of the road as one heads towards Longy, is a large boulder, on which is carved a cross and the inscription: 'E.C.L.W. 4th May, 1885'. It recalls that Captain Edward Charles Lethbridge Walter, of the 83rd (County of Dublin) Regiment, was killed there while driving his tandem.

Nearby, the road forks, one branch leading to Simon's Place, Whitegates (former Army married quarters) and the Fort Albert area, the other pointing to Longy Bay, but a better approach to the beach is to walk down Essex Glen by a charming little track formerly styled 'Barrack Master's Lane'. This takes one almost to the Nunnery. Alternatively, there is a path running south to Essex Castle, whence a rough road runs down to the shore.

Until the building of the breakwater and forts, Victoria Street was of small importance; today it is Alderney's real 'High Street'. Here are the town's principal shops, though private residences are not wanting. There are places of refreshment, too, including the Chez André Hotel, occupying the site of the Post Office, now in High Street. The Victoria Hotel is lower down the hill and on the other side of the road is the former Riou's Hotel. Off Victoria Street is Ollivier Street, where the Militia Arsenal once stood. On the corner is Albert House.

Facing Ollivier Street is the Albert Memorial gateway, a handsome granite arch with fine wrought-iron gates, recalling the visit of the Prince Consort. This is the principal approach to St Anne's Church, though there are others in New Street (formerly Rue des Heritiers), where the 1845 Court House and gaol stand, and in Le Pré, the hill running behind the church down to the Valley.

Had it been standing on a higher, more commanding site, St Anne's Church would have looked even more impressive than it does. It is built on the slope of a shallow valley, whose opposite side is graced by the Terrace, a small wooded pleasure ground, once forming part of the Government House property and today rather neglected. The church, even so, is very fine. Its warm Alderney sandstone and white Caen stone, the Norman/Early English style and the well kept churchyard backed by trees, create a delightful picture and the senses are even more gratified when its bells are ringing.

It was the gift of the Rev John Le Mesurier, son of the last of Alderney's Governors, in memory of his parents. Sir George Gilbert Scott designed it and the consecration took place in 1850. It is large, but then it was intended for the garrison as well as islanders, and it is, of course, the only Anglican church in Alderney. Within, it is most impressive, with its nave of eighteen arches, an apse wherein stands the high altar and beside it the Lady Chapel. Altars stand in the north and south transepts and, looking round, one finds it hard to believe that during the German occupation the building was so hardly used. Postwar restoration work has been well done.

The bells have been recast and rehung. The postwar windows are especially remarkable: one, in the Lady Chapel, commemorates the royal visit of 1957; another recalls the work of the services during World War II, and island scenes enhance its design; a third, near the war memorial, has Flanders poppies as its *motif*; and yet another remembers folk of other lands. There are a few Victorian specimens remaining. One is above the high altar, and below the conventional designs a keen eye may detect a chalice and patten etched in the glass.

Butes

Farther down Victoria Street is the island war memorial, standing in a small garden. It was first unveiled in 1923 and the names of the dead of World War II added later. At the foot of the street is a dwelling, now occupied by the resident Roman Catholic priest, which was once the home of Judge R. W. Mellish. It adjoins Les Rocquettes, a short stretch of road linking Victoria Street with Braye Road and the foot of Le Val. Close by is the lofty house of Val des Portes, once the residence of Judge French.

Butes has a place in the island's annals, for it was a place of assembly. Some of the events of Alderney Week take place there, and military reviews were once held there.

From the northern end, in particular, the prospect is vast. One can watch the never-ending procession of ships proceeding up and down the English Channel, Braye lies at one's feet, westwards are Burhou, Ortac and the Casquets and eastwards is Mannez. Some spectacular sunsets can also be enjoyed from Butes.

Various paths take one down to the shore and, while the main road certainly leads from St Anne to Braye, it is more pleasant to follow the serpentine path from Butes. It winds down the hillside to join Braye road not far from a diminutive lighthouse, one of the marks shipping use to enter harbour.

BRAYE

The port is severely practical, though the row of eighteenth-century houses in Braye Street is agreeable. The old pier, dating from 1736, once accommodated the Le Mesurier privateers, but more recently sailing ships discharged coal and loaded stone there. Today is it rarely used, though fishing craft and pleasure boats moor close by. The 'lower deck' of the pier is called 'Douglas Quay', and an inscription, carved in sandstone, confirms this on its façade. It was named after a former Lieutenant-Governor of Guernsey, Sir James Douglas, who held office in 1840, the date on the stone.

The jetty, now used by most ships calling at Braye, was completed early in the present century, after a good deal of controversy as to the wisdom of building it. Formerly, apart from the old pier, vessels had to use the breakwater slip or, when the tide served, the New Harbour.

Powerful cranes were set up by the Germans on the ugly extension they built, but they were found to be useless after the war. The extension itself was used by excursion ships and war vessels on occasion, but its condition began to deteriorate and it was closed. The States, whose unwelcome property it became, resolved to demolish it, but such a massive undertaking would be costly, and to date the structure has been left alone, yet another unsightly echo of the unhappy war years.

For something like half a century a forbidding looking building known to all as 'the crusher' dominated the Braye scene. Here stone was pulverised for shipment, and during the occupation the building was reinforced and adapted by the Germans for defence purposes. Part of it was demolished after the war (by which time it had become useless, since the granite industry was not revived), and in 1972 it was laboriously razed.

In Braye Street stand the Sea View Hotel and the Divers' Inn, both establishments of age and character. The former stages many social functions, while the latter, still retaining an engaging Victorian atmosphere, once accommodated none other than John Wesley, who slept there in 1787, following the introduction of Methodism to Alderney by Dr Adam Clarke. Wesley had not intended visiting the island, but his ship, bound for Guernsey, put into Braye owing to an adverse wind. He preached on Braye beach before sailing to Guernsey, and as a result Methodism took strong root in Alderney, as it did in the other islands. The first chapel in Alderney was opened in 1790, and its successor was built in Church Street in 1814.

The New Harbour looks rather unattractive when dry, but when it is filled with water, the picture is just the reverse. All manner of small craft, including fishing boats, ride at moorings and on the quayside is much gear and a useful marine store. Guns serve as bollards, and the harbour is dominated by Fort Grosnez and, to a lesser degree, by the former harbour office. The modern counterpart of the latter is a tower-like structure, beside which are the Alderney Sailing Club's premises. The stonework of the New Harbour was well wrought and the general effect is far more pleasing than the severely practical jetty at Braye.

The 'Couriers' and Other Ships

At the New Harbour the RMS *Courier* used to coal on occasion. At one time there were two *Couriers*, both plying between Guernsey and Alderney. Since they both had the same name, they were differentiated by being called the 'Big' and 'Little' *Couriers*. They were owned by the Alderney Steam Packet Co, which was formed in 1877. The 'Little' *Courier* was the first of the pair to be built—at Northam, near Southampton, by Day & Summers in 1877—and she had a gross tonnage of 105. Her sister, of 151 gross tons, came from the same yard and joined her in 1883. Both were fast elegant ships, each with a slim buff funnel, slender masts (on which sail was often hoisted) and a reasonable amount of passenger accommodation and cargo space. They were good sea boats, braving terrible weather at times.

The 'Little' *Courier* was sold in 1913, but the larger steamer remained in service until 1940, and, after working for the Admiralty during World War II, resumed her local sailings for a brief period in 1947, after which she was broken up. Long before then the Alderney Steam Packet Co had ceased to exist.

Braye, in former days, was often visited by War Department vessels, by Trinity House ships (which, like warships, still call there), by steamers which anchored in the roadstead for the purpose of trans-shipping explosives and, after World War II, by the *Island Commodore*, *Commodore Queen* and *Alderney Trader*, all local craft.

A REASONABLY good walker could cover the coast of Alderney in a day without effort, but unless his time was limited there would be little object in such an exercise. Better by far to explore the island's shore a little at a time, savouring the contrast between its low-lying region and its steep cliffs, and pausing to gaze at some of the finest coastal scenery the Channel Islands can offer.

STARTING AT BRAYE

It is as well to follow the coast clockwise, and what better starting point than Braye? The bay is Alderney's largest and its expanse is well seen from the old jetty. At low water there is a considerable stretch of sand, with rocky outcrops in the centre. Beyond looms Mont Touraille, a fine backcloth, with the rugged Roselle Point as an outlier. The bay has been artificially enlarged by the breakwater; formerly its western end was bounded by Grosnez Point, an area now absorbed by the harbour.

Whether one strolls over the sands or follows the grassy verge bordering the Lower Road (*La Banquage*, where once seaweed was dried), the way is enjoyable. The hinterland is reasonably impressive, with the grassy slopes running up to Les Rochers, whereon stands the television mast. A new road, from Newtown to the many modern houses above it, is too straight and steep to be attractive. Better is the old way of *Valongis*. The high ground is, in a sense, relieved by the mighty

Page 125 (*above*) The first President of the States of Alderney, the late Captain S. P. Herival (*left*) aboard HMS *Grenville* at Braye; (*below*) Alderney's breakwater. There is an upper and a lower deck and a railway runs on the upper. From it trucks tip stone over the seaward side in an attempt to check the force of the waves on the mole

Page 126 Gannets on the Garden Rocks. These magnificent birds chose Alderney as their home in 1940 and have stayed there ever since, returning each spring and remaining until the autumn. Beyond is Trois Vaux Bay and the Alderney heights

gash of the Battery Quarry, which the passage of time and familiarity have made tolerable to the eye.

Like all Alderney bays, that of Braye has had its share of wrecks. In January 1911 there were two at the same time. The first was the steamer *Burton*, laden with Alderney stone, which struck a reef twice. She burned flares, as it was night-time, and the *Courier*, then at Braye, put to sea and found her at anchor. The *Courier* could do little until dawn, when she started towing the *Burton* towards Braye, the crew having landed by boat. But the laden steamer was well down by the head and she had to be beached. Heavy seas soon destroyed her. The second wreck, on the same night, was the *Charles Ellison*, a barge, also carrying a cargo of stone, which broke adrift from her anchorage in the roadstead and was stranded on the rocks near the old pier. She also became a total loss, though her crew were saved.

Yachts sometimes anchor at the eastern end of the inlet. Here are the remains of a slipway once used for the cartage of seaweed from beach to farm, and seats provide comfort. Looking back along the usually busy road, one notices the small trees planted along its verge, a successful attempt to introduce a touch of the exotic in an otherwise rather stern panorama.

From the main gate of Fort Albert one can enjoy a wide view across the bay to the harbour and beyond it. The breakwater's length can be appreciated, as can the defensive situation of the fort, enhanced as it was by Roselle battery below it and the formidable rocks forming the seaward bastions. The low-lying expanse towards Longy might have been a sheet of water, if the military pundits had had their way and dug a canal to isolate eastern Alderney from the rest of the island.

Saye

It is enjoyable to stroll round the flank of Fort Albert eastwards, and find at one's feet the silver-gold arc of Saye Bay, with its flawless sands. The hillside descends rather steeply to the shore, but an easy path runs over the grass. Beyond the bay are more forts, sands and a mass of rocks, and a fitting terminal in the pillar of the lighthouse.

Saye (pronounced 'Soy') is Alderney's most symmetrical inlet. The somewhat prosaic hinterland only serves to emphasise the beauty of the beach, for behind it are sand dunes, rather featureless farmland and the inevitable German bunkers, some of which have been converted into summer houses quite successfully. Almost in the bay's centre is a grassy islet, easily reached except at high tide, from whose rocks one may dive into deep crystal-clear water. Oddly enough, the small area between the islet and the headland on which stands Château à l'Etoc is forbidding in appearance, with savage rocks and absence of sand—in strong contrast to the rest of the bay.

Corblets

On the right of the fort is what is called Arch Bay, though this is really only the north-western end of Corblets. It is so called because an arch connects the hinterland of Saye with the sands of its neighbour, and this was used by carts hauling seaweed. Above the stone arch is the road leading to the fort, itself a fine ingredient of the scene.

In 1866 the French barque *Carioca*, bound from Havre to Rio de Janeiro with a crew of twenty-nine and a general cargo, struck the rocks east of Château à l'Etoc and was lost. Her cargo included pianos and the sands of Corblets were full of them! Her crew were saved by gunners from the fort. The gale which caused the wreck was responsible for two others, also French, whose crews were lost on the island's inhospitable shores.

Corblets is one of Alderney's most popular bays. Its sands are broad, surfing can often be enjoyed and the rocks enfolding it are interesting to explore. The handsome fort on its eastern side enhances its character and appearance and the greensward below its walls makes a good resting place.

Veaux Trembliers and Cats Bays

A short distance east of Fort Corblets and its surrounding gardens is the curious Rocque Bertram, freakish in appearance, tall and easy to scale. The shore between the fort and the lighthouse is extremely rugged, the beach comprising ridges of stone so up-tilted as to make sitting problematical. Still, the turf bordering the foreshore makes a good resting place, and the massive rocks just out to sea add a touch of majesty to the prospect.

The inlet of Veaux Trembliers is succeeded by Cats Bay and between them lies the 'Islet of Flowers' (see p 77), where the ship *Liverpool* was lost in 1902 (see p 142). While it is simple for an agile climber to reach the ruined Fort Les Hommeaux Florains on the islet, he must be prepared to swim across a concealed gully, though there is really little to see on arrival.

Quèsnard to Longy

Quèsnard Point limits the other side of Cats Bay, and on the headland, near the lighthouse, is another fort. Hereabouts are signs of the German occupation, and they must be accepted, for to destroy the massive concrete would be laborious and costly and it has now become part of island history. Less forgivable is wartime barbed wire, which is almost indestructible. It should have been pitched into a quarry at the liberation, instead of being allowed to disfigure more than one quiet inlet in this part of the island.

A particularly enjoyable walk can be taken along a path running from Fort Quèsnard to Longy. Here one can look across the uneasy Race to the French coast; on a clear day the cliffs of Le Contentin can be seen, the great atomic energy station on the heights standing out as prominently as the lighthouse on Cap de la Hague. Only small vessels bound for Jersey, Guernsey or St Malo traverse the Race. Inland, the country is undulating, with abandoned quarry workings, several modern houses and a low hill, La Grande Folie; beyond that is Mannez Garenne, rough ground adjoining Longy Common.

The islet and causeway of Houmet Herbé (whereon stands an attractive fort) enclose the inlet of St Esquerre Bay. On the fort's other side is Baie du Grounard. Both bays are extremely rocky, though one or two sandy gullies may tempt the bather. Out to sea the water is usually turbulent, thanks to the Brinchetais reef, where many a ship has met her doom.

Approaching Longy one comes across disused rifle butts, together with a length of ruined wall. War Department stones abound, arguing the former military importance of this low-lying region. An artificial causeway, covered at high water, runs out to Raz Island, formerly known as *Le Houmet de Longis*, whence one can enjoy the prospect of Longy Bay, Essex Castle and the Hanging Rocks beyond it, the rather desolate area inland of the bay and the high ground extending from Essex Hill westward. Inland, one sees the summit of Fort Albert, though it merges so well with the landscape that it is by no means prominent.

Longy Bay

At the eastern end of Longy Bay, near the causeway, is a bed of peat, suggesting the presence of trees in ages past. The German anti-tank wall, though unattractive, has its uses, for it protects the hinterland from blown sand and makes a wind break for those reclining on Longy's broad sands. The water is relatively shallow, and at low tide it is a long walk to the sea. The bay is large, quiet and delightful. Walking across its sands is the best method of reaching its western end. Here is the Nunnery, some of whose ruined walls lie on the beach. A modern concrete slipway has replaced one made of stone and used by generations of fishermen and seaweed gatherers. A boat or two is kept here in summer, but Longy has long ceased to be of marine significance, though the few remains of its old pier can be found below Essex Castle.

A path runs from the Nunnery to Frying Pan battery, from which one gains an unusual prospect of the Hanging Rock, looming above. A former military building is now the Barn Restaurant. Another serves teas.

In 1826 the mail cutter *Hitchinbrook* was wrecked at Longy, though her passengers and mails were saved. In 1910 Raz Island causeway was the scene of the stranding of the Spanish steamer *Felix de Absalo*, in thick fog. Two women living nearby heard cries and notified the coastguards, whose quarters were close to Longy. (The buildings, of brick, are still called 'Coastguards'.) Troops from Fort Albert joined in the rescue operations and all were saved, including the captain's wife and her baby, born the previous day. All were landed at Raz Island. Coal and tobacco formed the main part of the cargo, islanders noticed.

ALONG THE CLIFFS

The cliffs extend from Essex Hill to Hannaine Bay. In places their dignity is slightly softened by the presence of inlets and sandy stretches, but in the main these heights are sheer and cruel in aspect. One may walk their summits with ease, but only here and there are descents to sea level possible.

A stony road runs up from Longy to Essex Castle. At the top is the site of an early radio-telegraph station, more recently used by the States of Guernsey Telephone Council. A rough track leads seaward to the Hanging Rocks—masses of granite leaning, like Towers of Pisa, at an acute angle over the cliff side. One is far larger than the other and neither is easily climbable.

Down through gorse, heather and brambles runs the path to the inlet of La Tchue, a curiously named bay which cannot be reached without a severe scramble. The track leads on to the cliffs of La Haize, where the rubbish of Alderney is dumped. Lorries use a track running from the Longy road to the cliffs, and while most of their contents falls into the sea, some, unfortunately, adheres to the vegetation. The rock formation here is fine, and it is a pity that a less primitive method of rubbish disposal cannot be found.

La Cachalière

The Blaye borders the path inland and from time to time small valleys and their streams run into the sea. One of these is the Vallet au Fleaume, between the Haize and Les Becquets headland. Soon the mass of L'Etac de la Quoire can be seen, standing offshore and accessible at low water via a path running down to the ruined Cachalière pier, beside which is an abandoned quarry. Stone from here was loaded into ships berthed at the pier, but the enterprise lost favour after the steamer *Tyne* was lost soon after loading there. The pier became known as 'Chicago', because Matthew Rowe, who is supposed to have made a fortune in that city, spent it on this pier, which proved a failure. The path leading to it also takes one to 'Bluestone Bay', roughly east of the pier, but it can only be reached when the tide is low.

Val du Sud

The cliffs can be reached from St Anne by several tracks (*routes de sufferance*) crossing the Blaye. One runs from the top of Le Val to above Cachalière. Another, from the Brecque, takes the walker to Val du Fret, west of Cachalière, while yet one more, from Little Street, skirts the airport and ends at Val du Sud. The cliff path passes the modern house called 'Quatre Vents'. At Val du Sud there is a pleasant diversion, thanks to the good taste of the States of Alderney. Trees have been planted, much undergrowth has been cleared, the path has been restored and the beauty of this little valley can be appreciated to the full. Access to the sea, however, is denied to all but the intrepid. The path rejoins the main one, which soon leaves the cliff edge, and the walker no longer enjoys such fine prospects. Still, with a little perseverance, it is feasible to scramble westwards slightly below the clifftop, and find many flowers, rock formations and

132

glimpses of the shore. At Val de l'Emauve are rocks that once formed the so-called 'Lovers' Chair', partly destroyed during the occupation.

Telegraph Cliffs

Another vale, Vallée des Gaudions, runs seaward until one reaches a headland, near Telegraph Tower, overlooking two of Alderney's finest rocks—La Nache and La Fourquie—marking the eastern limit of Telegraph Bay. From a former German bunker one can gaze down on these mighty stacks, 100ft or so high, though a still better view of them is to be gained from the bay itself.

Between this point and Tête de Judemarre the cliffs are at their finest—lofty and steep, yet softened by heather, broom, gorse and other vegetation. Below these forbidding heights are the sands of Telegraph Bay (anciently *La Foulère*), so called because of the tower above it and not on account of the telegraph cable which once ran to Guernsey. One may discern the ruins of a hut, from which strands of cable are visible. From there it ran down the cliff, under the shingle and out to sea.

Telegraph Bay

Perhaps the finest view of the bay is had from its western end, at Tête de Judemarre. From here one can see the sands— if the tide is low—the perpendicular cliff of golden hue, the great rocks, and, as a single contribution from man himself, the pleasing Telegraph Tower, solitary on the skyline.

Descent to the bay is not difficult to those prepared to face the rather laborious return journey. There are steps and a path, constructed about 1900. Those wishing to explore the bay should observe the tide, for at high water it reaches the foot of the steps, and if one is at an extremity of the beach, one cannot reach them.

Bathing here is excellent and the cliffs provide good shelter from the wind. The beach is well worth examining, for its stones are beautifully coloured, there is a small cave below the telegraph hut and the eastern end has fascinating rock pools. The presence of La Nache and La Fourquie, however, most impresses one, and they have, as neighbours, two lofty pinnacles known as 'The Sisters'. The western end has majesty, too, and is a good place for fishing.

Trois Vaux

Many leave Telegraph Bay by a track over arable land leading to the road. It is preferable to follow the path to Tête de Judemarre and then continue along an admittedly sketchy trail to Trois Vaux. Two small valleys merge into a third, and it is simple to scramble down to the brook and follow its course to the brink of a low cliff. Here the walker is advised to stop, for the face is crumbly. Below is the bay of Trois Vaux and out to sea are Les Etacs, which are often styled the 'Garden Rocks', perhaps because they are like guardians standing sentinel at the approaches to the Swinge.

One may clamber up the central valley, but it is hard going. It is rather easier to climb the steep northern slope and gain the track on another striking headland—La Giffoine.

La Giffoine

Massive remains of German defences fail to rob this eminence of its dignity. Here, long ago, stood one of Alderney's beacons. The view is superb, and today the headland forms the best vantage point from which to see the gannets on Les Etacs, short of visiting the rocks themselves. The steep cliffs are usually gay with flowers, and with heather in autumn. Gulls and other wildfowl abound. Across the water lies Burhou, beyond is Ortac and farther off the white buildings on the Casquets.

Progress along the clifftop is far from easy, since the path has been allowed to become overgrown. Probably it is wiser to take the rough road leading inland to the tarmac highway, though in so doing one misses a rather attractive little valley above Hannaine Bay. The high road may be followed until just short of the junction with the road running down to Fort Tourgis. At this point a small building will be seen on the left, and beside it a path, which should be taken.

Clonque

There was a time when this grassy path was marked 'Foot-path to Clonque', but the sign (like others in Alderney) has long since disappeared. The path soon begins to descend, zig-zag fashion, to sea level. Seats are plentiful and it is wise to pause at the top and savour the prospect below. Fort Clonque is the centrepiece and its backcloth is the Swinge. The bay and much else spring into view, at all times looking attractive from this vantage point. The name 'Clonque' is believed to be derived from the French *calange* (rocky inlet), which patois turned into *calanque*, modernised later to Clonque.

The zigzag provides the easiest of descents to the shore. Near the bottom is a piece of wall, probably all that remains of the barracks once serving Fort Clonque. A road of sorts leads to the causeway, and this is used by vehicles servicing the building and by those who spend out-of-the-ordinary holidays in its flats. Fort Clonque stands on an islet and from certain angles (especially from neighbouring Hannaine Bay), one may spy an aperture in the islet, through which daylight glimmers.

Hannaine Bay provides moderate bathing. Tongues of sand lie between masses of rock, but its most remarkable feature is the fine raised beach and strata visible in the sheer cliff face. The Clonque littoral is of interest to the marine zoologist and to all who delight in roaming the shore. The so-called 'Monk's Chair' is situated near one of the slipways running down from the coast road.

Midway along the road is Clonque Cottage. On a spring night in 1922 those living there heard their dog barking, and thought intruders were abroad; but the sounds came from shipwrecked mariners. Their vessel was the steam coaster

Emily Eveson, of 360 gross tons, ashore nearby and laden with coal from Rouen. Her crew had taken to the boats in dense fog, and the dog's barks guided them ashore. The ship broke up, but her boiler survives on the beach, a reminder of yet another Alderney wreck.

Inland is the Pommier valley, but it is difficult of access. The coast road runs beside high ground nearly as far as Fort Tourgis. Here is a lane running from Clonque Cottage to the rear of the fort, but the walker is better off if he pursues the coastal route, taking him past a German bunker and so round to Platte Saline. At this point the fort looms above in striking fashion. A good road climbs the hill and another, running inland, passes the remains of a watermill before it ascends Le Petit Val and takes one to St Anne.

Platte Saline

Yet the coast still beckons, and the walk along the track past Saline Bay is only excelled by a tramp over its vivid shelly shingle. The sea has a habit of pounding heavily on this shore, often raising surf and a roar as the shingle is assaulted. Fort Platte Saline should enhance the prospect, but it has been so industrialised that it is almost unrecognisable as a defence work. The gravel works have nearly absorbed it, yet fail to rob Platte Saline of its appeal. The word 'Saline' suggests a salting, while 'Platte' means the flat area inland from the bay.

Fort Doyle once guarded the bay's eastern end, but today it is neglected, part of the picture though it is. New dwellings are to be found hereabouts, both at Platte Saline and Crabby. A narrow path passing Fort Doyle is all that remains of a railway cutting, an echo of the days when a mineral train ran to Fort Tourgis while it and its neighbours were under construction.

Crabby

Crabby Bay wears a sad look, for its sand is grey, perhaps because stone dust from the neighbouring York Hill quarry was

136

dumped there when the quarry was being worked, though one would have thought that by now wind and sea would have removed it. The bay is otherwise pleasant enough, with a good shape, fine rocks and the *glacis* of Fort Grosnez as its neighbour.

The scar of York Hill quarry is not obtrusive, nor is the nearby electricity works, except perhaps for the hum of its motors. The harbour area is near, and above it is Fort Grosnez, lending a strong touch of character to a prospect which industry and urbanisation cannot spoil. An agreeable way to return to St Anne is the Valley running up from Crabby. It is well wooded, its houses and gardens are charming, and it ends at the Terrace, near one of the many disused drinking fountains still existing in Alderney.

13   LIGHTS AND WRECKS

FROM time immemorial the reef of Les Casquets has lured mariners to their doom, lying, as it does, about 7 miles west of Alderney and almost directly in the shipping lane. Because of its ill reputation, Thomas Le Mesurier, Lieutenant-Governor of Alderney, in 1709 petitioned the Queen in Council to have lighthouses built there and this was granted.

It took very much longer before another hazard, the eastern point of Alderney, was similarly dealt with by the erection of a lighthouse at Quèsnard Point nearly 200 years later. In a sense, though, both these beacons were unwelcome for, as E. A. Martin wrote in his *History of Alderney* in 1810, 'before these lighthouses [Les Casquets] were erected, there was scarcely a gale of wind without some vessels being driven on the island and destroyed; by the wrecks the inhabitants were furnished with many valuable articles and with the timber they built their houses, many of which are still existing'.

LES CASQUETS

The rocks, whose name means 'cascades', were leased to a man named Le Patourel (a Guernsey name) for no less than 60 years. In 1785 the Governor of Alderney, Peter Le Mesurier, became the lessee. Le Patourel (or his servant) had the unenviable task of keeping a coal fire burning as a warning to passing ships. The flames in his armourer's forge (as the installation was described) were fanned by bellows. Fuel was probably taken from Alderney.

In 1785 three lighthouses were built—Donjon, on the north-

138

east, St Peter and St Thomas. In 1790 a number of revolving lamps were installed, an arrangement superseding the oil lamps in a copper frame which had replaced the coal fire. In 1855 the lighthouses' height was increased, and by 1877 only one lighthouse was kept going, the others being shortened and employed for other purposes.

For 18 years during the nineteenth century one family lived on Les Casquets, 'without ever leaving the spot', wrote Mrs Lane Clarke in her *Alderney Guide*. They bore the island name of Houguez, and apparently the whole family, not just the keeper, kept watch by turns. They grew a few vegetables in earth imported from Alderney in what little shelter could be obtained, and to this day the remains of the garden can be seen. A boat supplied them with necessities once a month, otherwise they were isolated from the world. On one occasion a young carpenter visited the rock to effect repairs and fell in love with a daughter of the family. He persuaded her to visit his native isle of Alderney and she did so, only to return to the peace and quiet of the Casquets, for the island was too noisy for her! Nevertheless, she married the man and settled in the island. She was later joined by her father, Lucas Houguez, who was pensioned off when he lost the use of his limbs as a result of his long confinement in the damp conditions of the lighthouse. This family is mentioned by Swinburne in his poem 'Les Casquets'.

Alexander Deschamps' *Sailing directions for Guernsey, Jersey*, etc, first published in 1806, states that at Les Casquets the Governor of Alderney had 'a handsome and convenient house and visitors are treated with hospitality and kindness'. The keeper also had a comfortable house. It is hard to believe that such a remote place, often so difficult of access, should receive many callers, yet more than one writer has described visiting the rock. According to Jacob, the rocks are so steep that 'a line of battle ship may lie alongside them'. F. F. Dally, writing in 1863, stated that a storm completely destroyed the lights in

139

1823. Deschamps stated that supplies were only delivered in summer, because winter landings were too dangerous, though Mrs Lane Clarke, writing a little later, inferred that the service was kept up throughout the year. Visitors to Les Casquets now must first obtain written permission from Trinity House.

This corporation assumed control of the lights about the middle of the nineteenth century. It had been aware of the reef's menace as far back as 1709, for Thomas Le Mesurier's petition to Queen Anne stated that among those who went to view the rocks were 'the gentlemen of the Trinity House', as well as the Lord High Admiral himself.

Originally, of course, sailing boats from Braye took out keepers and supplies to the Casquets. For several years during the early part of the present century the MV *Lita*, a pilot craft, performed this duty, and from 1946 the MFV *Burhou* effected relief and maintenance duties. But in 1972 Trinity House changed over to helicopters, which now transport keepers, workmen, fuel and food to the Casquets and to the Alderney lighthouse, specially prepared pads being used at both stations. The virtue of these aircraft is in their operation when heavy seas make a boat landing impossible. If a landing cannot be made on the 'helipad', supplies can still be winched down. Another advantage is that keepers are no longer likely to be marooned on the rock for long periods because of adverse weather, and this will result in more regular reliefs. Doubtless very heavy building materials will continue to be taken over by boat, as in the past.

Passing reference has already been made to the 1942 Commando raid on Les Casquets, which the Germans were using as a naval signal station. The lighthouse was also attacked by the RAF and the lenses were badly chipped.

When access was by the tender *Burhou*, she used to anchor off and ferry supplies by dinghy to one of the three landings, of which the southerly was the most used. A long flight of steps leads up to the buildings, which are enclosed by a stout wall.

Within are the three towers and the living quarters—stone-built cottages and very comfortable.

Because of the spacious area of rock, conditions are far less cramped than at, say, Les Hanois, off Guernsey, whose tower stands on a relatively small area of rock. There is, at the Casquets, a yard in which to stretch one's legs, the remnants of a garden and enough room to fish from several vantage points.

The light is housed in St Peter's tower, standing 120ft above high water and with a visibility of about 17 miles. Electricity is supplied by a generator, installed in 1953, when oil lighting ended, and the light is of 2,850,000 candle power. St Thomas's tower houses the charging plant for the radio beacon, whose lattice masts form a conspicuous feature on the rock, and the electric lighting plant for the buildings. The Donjon tower, which contains the powerful fog signal, stands next door to the old Bell tower, where once a warning bell was tolled.

It was long believed that the celebrated White Ship struck the Casquets in 1120, but she probably foundered near Barfleur. Aboard her was Prince William, only son of King Henry I. One of the Casquets rocks is named 'White Ship' on the supposition that the vessel was lost there. On firmer ground (though even this belief is not completely proven) is the wreck of HMS *Victory* on the Casquets in 1744, with the loss of 1,000 lives, including Admiral Sir John Balchen, whose flagship she was. There were no survivors.

The loss of the Southampton mail steamer *Stella* at Easter 1899 is still remembered locally. The disaster occurred in thick fog and the ship, after striking the Black Rock (part of the reef), sank quickly, with the loss of 102 lives out of the 200 aboard. Some of the survivors were picked up by the Southampton mailboat *Vera* and the Weymouth packet *Lynx*; others were landed in Alderney. Among those drowned were the master of the *Stella* and a stewardess, Mary Rogers, who gave her lifejacket to a passenger.

ALDERNEY LIGHTHOUSE

The *Liverpool*, Alderney's most famous wreck, went ashore near Fort Hommeaux Florains in February 1902. Her 6,000 tons of cargo, including cement, girders, coke, marble and much foodstuffs, silks, soap and candles was salvaged both officially and unofficially (by the islanders). Some of the marble found its way to Guernsey and other salvaged material was taken there by the small steamer *Pioneer*, owned by a syndicate who bought the wreck for £250. For some time the proud *Liverpool*, with sails still set on her four lofty masts, became one of the wonders of Alderney. There were trips from Guernsey to view her and attempts made to save her.

By June 1902 she was still hard on the rocks and high tides covered half her decks. Many of her fittings were removed that summer, but hopes of saving her hull faded. Local museums have souvenirs of the ship, and until the German occupation some of her cargo was in many an island home. In the 1950s a party of visitors scrambling on the rocks near the derelict fort found some tins of sardines, and these, with their very old-fashioned appearance, probably came from the wreck.

Everybody refers to the lighthouse as 'Mannez' (pronounced 'Moanay'), yet within the building is a plaque stating that it is Alderney lighthouse. There is no other in the island, unless one includes the two diminutive robots, one on the old pier, the other at Braye Gates, at the foot of the hill, both of which lights, when kept in line, guide vessels into harbour after dark.

The *Liverpool* was by no means the first ship to be stranded at this end of the island, for in 1895 the steamer *Behira* had sunk there and in 1905 the SS *Portsea*. Many other vessels had also been wrecked in Alderney waters because there was no warning light at the junction of the Race and the Swinge.

In September 1910, therefore, Trinity House invited tenders for the construction of a lighthouse, and that submitted by

Page 143   (*above*) Ortac, a stack standing between Alderney and Les Casquets. Gannets breed on this mighty rock which is not difficult to scale, and the birds do not seem to resent visitors; (*below*) a striking aerial view of Les Etacs, or the Garden Rocks. The summits are covered with birds, while others hover over them. The rocks stand at the western end of the island

The great rocks of La Nache and Le Fourquie stand in Telegraph Bay. Steps leading down to the sands are to be seen in the foreground. The cliffs are the finest in Alderney and the bay is its most picturesque

William Baron, of Alderney, was accepted. By 1912 it was completed. In the previous year the Alderney States had granted a petition that a strip of land formerly used for the drying of seaweed be ceded to Trinity House, and that the former coast road be deviated behind the lighthouse property.

As at Les Casquets, the living quarters, stores and machinery are situated in buildings adjoining the lighthouse, which is a hollow structure, at the top of which is the lantern chamber. Stone for the buildings was quarried nearby. The tower is handsome, and wider than average, and from its gallery there is a broad view.

The lamp has an intensity of 400,000 candle power, with a range of about 17 miles. The fog siren is situated in a turret off the engine room, and outside the tower are its two big trumpets. Like all lighthouses, the property is immaculately maintained and, fortunately, it was not damaged during the occupation, when the Germans occasionally used it.

BURHOU

The islet of Burhou lies just across the Swinge, and remains the Channel Isles' sole 'desert island'. True a wooden hut there can be rented, if application be made at the States Office, Alderney, but the real residents are the birds, and man is there only on sufferance. If he does elect to stay for any length of time, he should take food and water, for there is no drinkable supply.

Burhou is 1,200yd in length and nearly 400yd wide, the highest point being 83ft above high water. A reef extends $\frac{1}{2}$ mile west of Little Burhou (the western end of Burhou) and it is covered at high tide, apart from one or two rocky heads. Between this reef and rocks just east of Ortac is a navigable channel and another, La Passe de la Maure, north of the reef, leads into the Swinge. Obviously, no amateur yachtsman in his senses would attempt these passages without the aid of a pilot.

The Renonquet reef lies north of this 'Passe' and it is also about 1,200yd long. The western end is immersed at high tide, but lofty rocks at the east are never covered. Just north-east of Burhou are the impressive Nannels, the highest rock rising 6oft above high water. Again, this is a hazardous region, not to be attempted by the tyro sailor.

The Swinge is aptly described in *The Channel Pilot*, which warns of the existence of dangerous rocks and points out:

> . . . there is much broken water, even in the calmest weather, caused by the rapidity of the stream over the rugged bottom. There are also two overfalls, caused solely by the tide, of which the danger is much increased in bad weather. During easterly winds, on the north-eastern stream, overfalls extend nearly across from Burhou island to the outer end of Alderney break-water. With westerly winds, on the south-western stream, they extend from Ortac to the Etacs.

On a wild day, to stand on high ground between Butes and the Giffoine is to behold a magnificent sight seaward, with the Burhou and Swinge area a mass of 'white horses'.

Those bound for Burhou will observe 'rafts' of puffins as they draw near. Formerly, this engaging little bird was more numerous than it is; the black-backed gull, however, is its natural enemy, feeding on puffins' eggs and chicks to such an extent that the smaller creature is being steadily exterminated. Yet many still remain and it is pure joy to stay quiet near their burrows in the island's friable soil and watch them emerge, peer about and perhaps fly off, to return later with a beakful of fish. Puffins also group themselves on rocks in picturesque fashion, and their appearance and almost human antics are alone worth going to Burhou to see.

The hut stands within the ruins of a stone cottage, built in the last century partly as a dwelling and partly as a refuge for ship-wrecked sailors. A supply of 'iron rations' was kept there for that purpose, a practice continued in its successor, which also contains a few rudimentary utensils, as well as bunk beds.

Seals are sometimes seen on the rocks and reefs off Burhou,

though there is no evidence that they breed there. They are regarded with some disfavour by fishermen, who believe that they, like the gannets and other seabirds, adversely affect local catches.

In his *Archaeology of the Channel Islands* Sir Thomas Kendrick quoted Holinshed's name for Burhou—'Ile of Rats'—'so called of the huge plentie of rats that are found there'. Holinshed stated there were so many that they drove out the population. There are no rats on Burhou today and one is tempted to wonder whether there was confusion with Ile de Raz, the old name for Raz Island, on the fringe of the Race. On the other hand, 'race' could well describe the course of the Swinge, which has also been styled 'Passage de Singe', or 'Monkey's Passage', though surely this is a fanciful translation.

Lukis thought some heads of stone found there might be of prehistoric origin, and in 1928 flint flakes, 'clearly the debris of a prehistoric chipping site', were seen by J. E. Ainsworth, who also noted 'two standing stones' that he took to be the remains of a circle. Geologists visiting Burhou in 1936 found pottery sherds and flint flakes.

Apart from shipwrecked seamen and naturalists, the only other human residents of the island seem to have been a French couple who, about 1900, lived in the cottage and kept a few goats and pigs. Dr Ludwig Koch made sound-recordings of island birds, and visited the gannetries with Charles Toms of Guernsey. The writer R. M. Lockley also spent two holidays on the island, the second in the 1950s, as he mentions in his charming book *Puffins*.

Birds are everywhere. Gulls of various types predominate, but as well as puffins there are razorbills and guillemots, shags (whose evil-smelling nests are to be found in the strangely contorted outcrops and oddly shaped boulders), and rock pipits. At night storm petrels haunt the ruins of the cottage. They used to nest in its loft, and even after the Germans, using it as a target, shattered it with gunfire from the Alderney batteries, the little birds did not forsake the site. They bred on Burhou long before the cottage was built, for John Jacob in his *Annals* described their breeding there.

Reference has already been made to a big Burhou and a little one, the two being separated by a narrow channel, dry at low water but often a torrent at other times. The vegetation of the smaller islet is quite different from that of the larger, for among other peculiarities it boasts a quantity of bladder campion not found on its neighbour. However, 'big' Burhou has much sea spurrey (rare on its companion), which thrives on a friable soil, rendered thus by the burrowing of rabbits and puffins. Bluebells and bracken grow well on Burhou, though it is innocent of trees and little scrub is seen. There are, in all, about fifty species of plants there, including grass, a newcomer.

Shipwrecks on Burhou have been very numerous. One of the most impressive was that of the big Turkish steamer *Edirne*, which stranded there in 1950. She was towed off, but foundered off Mannez. No lives were lost, unlike the wreck of the *Crispin*, a brigantine, which sank off the island with all hands in 1865. A year later the schooner *Jules* came to grief there, but her crew were saved by the steam tug *Watt*, used on breakwater construction work. A large German steamer, the *Leros*, was lost there in 1906, and in 1910 the SS *Linn o' Dee* ended her days on this inhospitable coast.

In 1897 the French SS *Marie Fanny* was lost in a gale and the master and his dog were the sole survivors. The weather had detained two sportsmen on the island and, returning in the darkness from rabbit shooting, they were amazed to find the Frenchman in the cottage. When the steamer *Rhenania* stranded there in 1912, some of the Friesian cattle aboard were landed on the island and the ship's mate stayed there for 3 weeks, tending them.

Possibly the most dramatic of the wrecks in these waters was that of the turbine-propelled TBD *Viper*, lost in 1901 during naval manoeuvres. She was among several ships of the Channel

Fleet operating in the vicinity of the Channel Islands and was part of a squadron based at Alderney. HMS *Viper* was of 344 tons displacement, with a maximum speed of 33.8 knots. She carried no radio.

In these manoeuvres two opposing fleets were contending for command of the English Channel and its approaches. The limits of Braye declared as being proof against attack comprised a line drawn from the submerged end of the breakwater to the northern extremity of Château à l'Etoc. Lack of visibility during these operations resulted in squadrons passing each other without interception, while spring tides added to the hazards.

On 3 August 1901 in dense fog, the destroyer was lost on the Renonquet reef, north of Burhou. There was no loss of life, however. Thirteen days later, her commanding officer, Lt W. Speke, RN, was court-martialled at Portsmouth and was found guilty of negligence and reprimanded. He admitted that before the disaster he had not used the deep sea lead nor the patent log, relying on the speed worked out by his engineer for accuracy. He did not verify his position, and only 35min before *Viper* struck he was in about 25–30 fathoms and 6–7 miles west of the reef. Distress flares from *Viper* were seen by the Alderney pilot cutter *Volage* and a destroyer, which went to the rescue.

Examination revealed that *Viper*'s hull was badly broken. Much of her armament, ammunition and stores were salvaged, but it proved impossible to recover her turbines. The destroyer was blown up by the Navy on 17 August and her remains were sold to a Southampton salvage company for £100. Much of the vessel still survives, to the satisfaction of local skin divers.

During the manoeuvres another casualty was TB 81, which struck the submerged end of the Alderney breakwater. With much difficulty she was brought to the newly built jetty, where stores, etc, were removed. The SS *Staperayder* was loading stone there, and she was placed on one side of TB 81 and two more torpedo boats on the other, with a chain slung under the stricken craft's hull. However, her weight nearly submerged the torpedo boats, so the chain was removed and she sank to the seabed. Later she was raised and towed to Portsmouth, surviving until 1921.

ALDERNEY is prosperous. In the main, this prosperity has meant that what has needed to be done has been done. For example, a new school became imperative. The old Le Mesurier building had served for years, and from time to time additions were made to the original fabric, but it was clear that the island teachers and their children were worthy of something better. So a fine new school was built near Newtown for children aged 5 to 15 on an excellent site and with sufficient room for future enlargement. The money was found and it was well spent.

The States of Guernsey Education Council controls education in Alderney, although a local committee exists, on which members of the Alderney States are represented. Alderney children attend Guernsey schools for secondary education and for this, and for higher education and training, grants are awarded. These are on the same basis as those of the United Kingdom. As well as the Alderney School near Newtown there is an excellent private establishment—La Brecque School. To some extent this replaced the Convent School, after the nuns had left Alderney and their house had become the Island Hall.

Again, a better drainage system was badly needed, road surfaces required improvement and a civic centre was wanted. These necessities were duly supplied, even if the island had to wait a long time for them. The postwar recovery period was lengthy and a sense of priority had to be maintained. Rome was not built in a day. In all matters involving capital expenditure, Guernsey was invariably sympathetic and any help needed, particularly of a technical nature, was gladly given.

Reference has been made to postwar commercial ventures, most of which came to an untimely end. It might be asked, how does Alderney live? What are its industries? They are not numerous. Tourism, some market gardening, the export of sand, gravel and grit, a very little farming and fishing—these are the chief sources of general revenue.

Much employment has been provided by the volume of building which has altered the face of Alderney over the postwar decades. This form of industry flourished with the 'invasion' of settlers. Just as, in the nineteenth century, Alderney was discovered through the construction of its fortifications, thus resulting in an influx of workmen, so, in the twentieth century, another colonisation has occurred, bringing new blood, new buildings, new ideas.

Since the 1840s, there has been a strong United Kingdom element in the population, and what are often regarded as typical Alderney names are, in fact, those of comparative newcomers. They have become so absorbed in the island's way of life that they have become accepted as fully as those families that have lived there for centuries, but which, in turn, were 'settlers' originally.

The postwar influx has done Alderney an immense amount of good. The new arrivals have been, for the most part, wealthy, middle-aged and quite capable of taking an active part in insular affairs. At the same time, it must be confessed that they have not brought children with them and there is a distinct danger that, with prospects rather poor, the youngsters of Alderney will seek work overseas.

Another disadvantage brought about by the postwar influx, it must be admitted, is that so many new houses have been built that the face of Alderney has been changed radically. Sometimes rashes of buildings occupy the sites of former fields or other open spaces, resulting in panoramas which, once wild or rural, now appear urban or suburban. In the main, the dwellings are attractive, but in the mass they rob the island of a measure of its appeal.

This fact caused consternation in Alderney. It seemed that a short-sighted 'get rich quick' policy might ultimately spell ruin to the community. True, the States Natural Beauties Committee exerted some control over building and there was a sacrosanct 'Green Belt', but there was a good deal of stifling of initiative in these controls and little vision.

THE DAVIDGE REPORT

A plan, a blueprint for Alderney's future was needed, and the States rose to the occasion in 1967, when Messrs W. R. Davidge & Partners, of London, were asked to submit a report as to the development of the island. Its author was E. E. Taylor, a principal of the firm. He had been asked to produce a development plan, bearing in mind the 'Green Belt', the fiscal link between Alderney and Guernsey, the availability of public services, the need to provide for a balanced population, the continuation of a prosperous economy, the conservation of the island's areas of outstanding natural beauty and the administrative difficulties of a too sophisticated system of planning control.

He reported that, of nearly 2,000 acres, 115 were used for urban and associated purposes, 460 for agriculture, 120 for airport and quarries, 90 comprised beaches and 1,135 acres were unused. The population was then about 1,780. He foresaw the conservation of what was scenically best, and the utilisation of virtually 'expendable' areas in building and commerce. He envisaged a development area limited, broadly speaking, to St Anne, Platte Saline and Valongis. He wanted the retention of the existing 'Green Belt', and advocated 'open areas' within the development area and a coastal footpath. He emphasised that it was neither desirable nor practical that building design be stereotyped, nor that too rigid control should

K

be exerted over details in designs, but, he wrote: 'It should be demanded that all new buildings and work to existing buildings is done to the design and under the direction of a qualified architect.' His report was accompanied by a 6in map, marking the areas under review.

Naturally it had its opponents, but when it was considered by the States in 1969, it was favourably received, though with some reservations, by Alderney's Island Economy Investigation Committee, which noted that there would be room for between 500 and 600 new houses, double the number then in being, which did not include dwellings for which permits had already been granted. The builders thought they could erect forty new houses a year, if they devoted their energies solely in this direction.

So far there is no marked evidence of the plan being adopted. Building still goes on, though the 'Green Belt' is, in the main, still untarnished. Clearly, the adoption of the Davidge Report will take time, but in the end Alderney will be vastly the better for it.

ASPECTS OF LIFE TODAY

As well as the two medical practitioners (of whom one is Medical Officer of Health, employed by Guernsey's Board of Health), there is the Mignot Memorial Hospital, which cares for the sick and aged. However, in the event of serious illness or major injury, patients are flown to either an English or Guernsey hospital. A reciprocal agreement exists between the United Kingdom and the Channel Islands, whereby visitors to either are financially covered in the event of sickness or injury.

The present Mignot Memorial Hospital, opened at Crabby in 1958, is managed by a local committee and is heavily subsidised by the States, though, even then, patients are expected to pay for treatment. The building was designed and paid for by the Nuffield Provincial Hospitals Trust and an island benefactor met the cost of building the matron's quarters, added later.

The St John Ambulance provides invaluable ambulance, first aid and rescue services. The fire brigade, an efficient body, is voluntary and is financed by the States.

On the business side Alderney has a flourishing Chamber of Commerce and a Royal Agricultural Society which is less active than in the past, when cattle breeding, exports and shows played such a role in the life of the island. Freemasonry thrives and the Temple is in Church Street. The Royal Antediluvian Order of Buffaloes has three lodges and is quite a feature of island life.

Animals are cared for at the clinic of the Alderney Animal Welfare Society in Le Val. The Women's Royal Voluntary Service is active, like the Women's Institute. There is also a fur and feather club. We have already dealt with the sporting and cultural sides of life.

The 'Aurigny' aircraft carry the emblem of a lion rampant, which is the device of the island. It is to be found in its official flag (the Union flag, with the lion as a centrepiece), on States buildings and elsewhere. It also appears on certain issues of Guernsey postage stamps, and a view of Braye harbour forms another stamp design. A former issue carried the exterior prospect of St Anne's church.

Those who live in Alderney may well be accustomed to its disfigurements, though the discerning visitor notices them at once. About the island he cannot fail to observe the unwelcome presence of derelict motor vehicles and disused farm equipment. Old iron is prevalent and too much untidiness stares one in the face. Frankly, Alderney is not well groomed in places.

This is especially noticeable at Braye, where an air of desolation prevails. True, the area of the stone crusher was daunting, and its removal and the development of the site will doubtless mend matters, but for too many years ruined buildings and a dismal atmosphere have formed a poor introduction to those coming to Alderney by sea, royalty included, to say nothing of those staying at Braye. Another defect is the almost complete lack of signposts. Certainly, one may buy a map (and the

Ordnance Survey is, of course, first class) and find one's way about, but the presence of even a few directions is, surely, an elementary requirement.

According to the 1971 census, Alderney had a population of 1,686, of whom 797 were males and 889 were females. This compares with a total of 590 for Sark (a slightly smaller island) and 51,458 for Guernsey. The census gave Alderney an acreage of 1,962, with 871 private households and 573 occupied dwellings.

It is interesting to compare Alderney's present population with that of some previous years. In 1821 it was 1,154. It rose to 3,333 in 1851 and as high as 4,932 in 1861, during the great wave of building, only to fall to 2,738 in the next decade (when much of the work was completed) and to 2,048 in 1881. The figure rose to 2,561 in 1911, but dropped to 1,598 ten years later. In 1939 it was 1,442 and in 1951 it stood at 1,328.

COMMON MARKET PROSPECTS

When it became apparent that Britain was about to seek membership of the EEC, the Channel Islands were perturbed. People feared loss of trade, or loss of privileges; and in some quarters there was even talk of independence from the United Kingdom if she became a member. Alderney shared in these apprehensions, and many were convinced that the island could run its own affairs independent of Guernsey and indeed of Britain itself.

A special States committee was formed to probe these possibilities and independence was very much in the air. The Home Office had stated that each island had the opportunity of acceding or not to the Treaty of Rome and, had Alderney taken a different decision than Guernsey's on the question of the Common Market, it would probably have meant the parting of the ways between the two islands.

In 1971 Geoffrey Rippon, MP, the Common Market

negotiator, addressed an extraordinary meeting of the States of Guernsey, at which Alderney and Sark were strongly represented. He stated the favourable terms offered to the Channel Islands, and this so impressed their representatives that the States of Guernsey and Alderney, with the Chief Pleas of Sark, accepted the proposals for entry into the EEC. At this meeting the President of the Alderney States, George Baron, thanked Mr Rippon for what he had achieved on behalf of the States. Mr Rippon, at a Guernsey press conference, declared his confidence that the island parliaments would accept the terms offered and, in due course, they did.

The Community's proposals meant that the Channel Islands' trading relationship with the United Kingdom would be safeguarded and, additionally, they would have opened to them new markets in the member states of the existing Community. The islands would apply the Community's external policies in relation to trade with third countries. For industrial purposes this would mean the application of the common external tariff to imports from third countries, and for agricultural products the application of the Community's system of levies, 'together with only so much of the internal measures of the common agricultural policy as will be essential to free trade'.

The islands' fiscal autonomy was guaranteed and value-added tax or any part of Community policy on taxation would not apply to the Channel Islands, Mr Rippon stated. All nationals of the enlarged Community were to enjoy equal rights, with no discrimination, but they would have no right to enter the islands to seek work.

Mr Rippon also said: 'I have managed to negotiate for you the maintenance of your traditional rights in the United Kingdom, where you will be able to compete for jobs on an equal basis with EEC nationals. But we who live in the United Kingdom will not retain our traditional rights of access to jobs in your islands.' He told his audience of a safeguard clause which would provide for the taking of *ad hoc* measures to meet any particular problems that might arise, but said there was no cause for anxiety over this provision.

He ended with these words: 'You will be free to maintain those features which today so strongly distinguish life in the islands from life in the United Kingdom.'

How Alderney and her neighbours will fare in the future is a matter for tomorrow's historians to relate. The prospects appear reasonably good, and prosperity, enjoyed so fully in the past, would seem to be assured. Alderney has experienced sunshine and shadow, good and evil, like all communities. She has weathered storms likely to have engulfed her, partly with the help of others, mainly through her own courage, fortitude, good humour and a firm refusal to be thwarted.

The Common Market is another feature along the road of progress. Some may style it an obstacle, others a milestone pointing the way to better things. Alderney has endured much. The German occupation was her greatest tribulation. Having overcome this, how can she be dismayed at anything which lies ahead!

APPENDIX A

Hereditary Governors of Alderney

1660 Edward and James de Carteret and Clement Le Couteur, of Jersey, before which time the island was 'farmed'
1682 Sir Edmund Andros, of Guernsey, who acquired rights from Sir George Carteret, of Jersey
1713 George Andros, of Guernsey
1714 John Le Mesurier, of Guernsey
1722 Anne Le Mesurier (*née* Andros)
1729 Henry Le Mesurier
1744 John Le Mesurier
1793 Peter Le Mesurier
1803 John Le Mesurier, who sold the patent to the British Government in 1825. Thereafter, Lieutenant-Governors of Guernsey assumed command

APPENDIX B

Judges of Alderney

1612	Pierre Herivel (first judge)
1637	Edouard Gauvain (in office at this date)
1641	Pierre Le Cocq
1653	Richard Le Cocq
1683	Nicholas Le Cocq
1702	Nicole Duplain
1703	Thomas Le Cocq (removed from office 1736)
1760	Jean Simon
1769	Jean Gauvain
1803	Jean Ollivier
1807	Pierre Gauvain
1837	Jean Gaudion
1856	Thomas Clucas
1876	Thomas Nicholson Barbenson
1892	Judgeship vacant
1893	J. A. Le Cocq
1898	N. P. L. C. Barbenson
1913	Major R. W. Mellish
1938	Brig F. G. French
1947	Sir Frank Wiltshire (last judge)

APPENDIX C

Presidents of the States

1949   Sydney Peck Herivel
1970   George William Baron

BIBLIOGRAPHY

ADMIRALTY. *Channel Pilot* (1906)
ALDERNEY SOCIETY. *An Alderney Scrapbook* (Alderney 1972)
ALDERNEY STATES. *Official Guide* (Alderney 1972)
ANSTED & LATHAM. *The Channel Islands* (1862)
CLARKE, LOUISA LANE. *The Island of Alderney* (Guernsey 1851)
COYSH, VICTOR. *Afoot in Alderney* (Guernsey 1969)
DESCHAMPS, ALEXANDER. *Sailing Directions* (Guernsey 1806)
DOBSON, RODERICK. *Birds of the Channel Islands* (1952)
HOME OFFICE. *Report of Committee of Privy Council* (1949)
INGLIS, H. D. *The Channel Islands* (1835)
JACOB, JOHN. *Annals of Guernsey, Alderney etc* (Paris 1830)
KENDRICK, SIR THOMAS. *Archaeology of the Channel Islands*, Vol I (1928)
LOCKLEY, R. M. *Puffins* (1953)
MARTIN, E. A. W. *History of Alderney* (Alderney 1810)
PACKE, M. ST J., & DREYFUS, M. *The Alderney Story* (Alderney 1971)
SHARPE, F. *The Church Bells of Guernsey, Alderney & Sark* (Brackley 1964)
TUPPER, F. B. *History of Guernsey etc* (Guernsey 1854)
WOOD, A., & M. *Islands in Danger* (1955)

NEWSPAPERS AND PERIODICALS

Evening Press & Star, Guernsey
Transactions of La Société Guernesiaise, Annual

ACKNOWLEDGEMENTS

TO many I am indebted for help and encouragement in the writing of this book. They include members of the States of Alderney, and especially the chairman and members of its Tourist Committee. I am grateful to Dr J. T. Renouf, Curator of the museum of La Société Jersiaise, for valued assistance in the geological section, and to La Société Guernesiaise, for permission to reproduce Dr G. H. Plymen's geological map of Alderney. Mr H. C. Timewell, of Guernsey, has given me additional information on the loss of HMS *Viper*.

To the authors of books and papers quoted in the text, and in particular to those published by the Alderney Society (whose members have given me unfailing support at all times) I am very thankful. My gratitude goes to the Alderney States Tourist Committee for sanction to reproduce the colour picture on the jacket, as well as the map of the island. My thanks go to the Guernsey Press Co Ltd for the picture of St Anne's Church and, not least, to Carel Toms, whose photographs enrich this book.

INDEX

Page numbers in italics refer to illustrations

INDEX